Empirical Studies on Volatility in International Stock Markets

Dynamic Modeling and Econometrics in Economics and Finance

VOLUME 6

Series Editors

Aims and Scope

The series will place particular focus on monographs, surveys, edited volumes, conference proceedings and handbooks on:

- Nonlinear dynamic phenomena in economics and finance, including equilibrium, disequilibrium, optimizing and adaptive evolutionary points of view; nonlinear and complex dynamics in microeconomics, finance, macroeconomics and applied fields of economics.

- Econometric and statistical methods for analysis of nonlinear processes in economics and finance, including computational methods, numerical tools and software to study nonlinear dependence, asymmetries, persistence of fluctuations, multiple equilibria, chaotic and bifurcation phenomena.

- Applications linking theory and empirical analysis in areas such as macrodynamics, microdynamics, asset pricing, financial analysis and portfolio analysis, international economics, resource dynamics and environment, industrial organization and dynamics of technical change, labor economics, demographics, population dynamics, and game theory.

The target audience of this series includes researchers at universities and research and policy institutions, students at graduate institutions, and practitioners in economics, finance and international economics in private or government institutions.

Empirical Studies on Volatility in International Stock Markets

by

Eugenie M.J.H. Hol
ING Group Credit Risk Management,
Amsterdam, The Netherlands

KLUWER ACADEMIC PUBLISHERS
DORDRECHT / BOSTON / LONDON

A C.I.P. Catalogue record for this book is available from the Library of Congress.

ISBN 1-4020-7519-7

Published by Kluwer Academic Publishers,
P.O. Box 17, 3300 AA Dordrecht, The Netherlands.

Sold and distributed in North, Central and South America
by Kluwer Academic Publishers,
101 Philip Drive, Norwell, MA 02061, U.S.A.

In all other countries, sold and distributed
by Kluwer Academic Publishers,
P.O. Box 322, 3300 AH Dordrecht, The Netherlands.

Printed on acid-free paper

Printed in the Netherlands.

Well, I don't think getting to the top is all that important. You can always have another go. The things you remember after a trip are not standing on the summit but what went on while you were on the route.

Mo Anthoine - mountaineer

Contents

List of Figures

List of Tables

Chapter 1

INTRODUCTION

Research into the behaviour of stock markets can be traced back as far as the beginning of the twentieth century when Bachelier (1900) introduced the random walk hypothesis for changes in security prices. Then, in the fifties, Markowitz (1952) developed what has become known as the modern portfolio theory, which basically states that in order to obtain higher expected returns one has to accept a higher level of risk. The importance of the variability of security price changes extends however beyond that of portfolio allocation issues as return volatility also plays a crucial role in a number of other areas such as the pricing of derivatives, hedging decisions and the calculation of Value-at-Risk measures. Perhaps its significance has been most concisely and persuasively summarised by Andersen and Bollerslev (1998) when they simply state

"Volatility permeates finance."

It is therefore imperative that we understand how asset return volatility, which is synonymous to the risk associated with the underlying asset, behaves over time. Return volatility cannot be observed and therefore it needs to be approximated. The most basic method to "calculate" return volatility is to define it as the average deviation from the mean value over a given time period which results in a constant. Since the early sixties it has however been recognised, most notably by Mandelbrot (1963) and Fama (1965), that volatility is time-varying, *i.e.* the degree of dispersion around the mean is not stable over time. Furthermore, the returns of financial time series exhibit prolonged periods of high and low volatility where shocks in the series are bunched together, as observed by Mandelbrot (1963):

"Large changes tend to be followed by large changes, of either sign, and small changes tend to be followed by small changes."

This phenomenon is often referred to as volatility clustering; volatility persistence is especially prominent in high frequency return series and this ensures that large shocks will influence future volatility for a long period after the actual event took place. In the past two decades two prominent classes of models have been developed to capture the time-varying autocorrelated volatility process: the Generalised Autoregressive Conditional Heteroskedasticity (GARCH) and the Stochastic Volatility (SV) model which are both defined by their first and second moments, also referred to as the mean and variance, respectively. The time-varying variance of the (G)ARCH model, introduced by Engle (1982) and Bollerslev (1986), is a deterministic function of past squared innovations and lagged conditional variances. The SV model, on the other hand, defines volatility as a logarithmic first-order autoregressive process with a stochastic component following initial research by Taylor (1986), Nelson (1988) and Harvey, Ruiz and Shephard (1994). The main distinction between GARCH and SV models is therefore that the latter has an additional error term in the variance. As volatility is a latent process, much of what we know about volatility has been learned from fitting these historical parametric volatility models and this has been complemented by the study of two alternative information sources: implied volatility and intraday volatility measures.

The main objective of this book is to investigate and analyse the Stochastic Volatility model in the context of international stock markets. Although SV models are generally recognised as a competitive alternative to GARCH models, their empirical application has been limited due to problems that arise as a result of the intractability of the likelihood function which prohibits its direct evaluation. Fortunately, in recent years considerable advances have been made in this area which has facilitated further empirical research and encouraged the development of extensions similar to those of GARCH models. Various of these SV model extensions are developed and their empirical implications are investigated for international stock indices. Their empirical performance is not only compared to that of the well-established GARCH model, but also to implied volatility measures and so-called realised volatility models where the latter are based on intraday volatility measures.

In the remainder of this chapter the research objectives which motivated the different parts of this book are discussed and an overview of the various chapters is given. It is noteworthy to mention that each chapter is self-containing and can be read independently as the research objectives underlying them originated at non-overlapping stages and attempted to answer different questions.

Chapter 2 presents the empirical stylised facts of stock index return series and gives an overview of the two classes of time-varying volatility models, the Stochastic Volatility (SV) model and the Generalised Autoregressive Conditional Heteroskedasticity (GARCH) model. The extent to which these models are capable of incorporating the salient features of stock index return series is also examined. These characteristics can then be summarised as follows: the volatility of stock index return series is not constant over time and returns experience prolonged periods of high and low volatility. In addition, the return distributions are leptokurtic. The main purpose of this chapter is therefore to introduce the general specifications of the SV and GARCH models and to discuss their empirical implications. In particular, it is found that SV models provide a "better" description of the salient features of stock index volatility than GARCH models.

Chapter 3 investigates the relationship between returns on stock indices and their volatility. For this purpose the so-called Stochastic Volatility in Mean (SVM) model is developed. This model incorporates unobserved volatility as an explanatory variable in the mean equation and can therefore be viewed as the SV counterpart of the (G)ARCH in Mean (GARCH-M) model of Engle, Lilien and Robins (1987). Our research is primarily motivated by findings in the GARCH in Mean literature which appear to indicate a positive relation between expected returns and expected volatility, which is however only sporadically statistically significant. As negative coefficients for the in-mean parameter are however also reported for GARCH models, the question arises which sign the in-mean parameter would take when the issue were to be investigated with SV models. The main difference between the two classes of models is that the GARCH-M model intends to estimate the relationship between expected returns and expected volatility, whereas the aim of the SVM model is the simultaneous estimation of (1) the ex ante relation between returns and volatility and (2) the volatility feedback effect. This fundamental difference implies that the estimated in-mean parameters of the two models have to be interpreted accordingly. An empirical study is presented in which results obtained with the SVM model for three international stock market indices are discussed. The selected indices are the Financial Times All Share (United Kingdom), the Standard & Poor's Composite (United States) and the Topix Index (Japan). The SVM model results are then compared with those of the GARCH-M model, which is especially interesting as such a comparative study has not been undertaken before. Whereas the in-mean parameter takes a positive sign in the case of the GARCH-M model, the SVM model equivalent is found to be negative.

The high degree of intertemporal volatility persistence observed with the time-varying models presented in Chapter 2 suggests that the variability of stock index returns is highly predictable and that past price observations contain valuable information for the prediction of future volatility. In Chapter 4 the forecasting ability of the historical-based volatility models is discussed. We show how the forecasting formulas of the different models evolve over time concentrating on those of the SV and GARCH models and present a research review. In the empirical literature contradictory results have been reported and hence there is still no consensus as to which model provides the most accurate stock market volatility forecasts. However, thus far only two of these studies have investigated the forecasting performance of SV models. Heynen (1995) and Yu (2000) concluded that the SV model had the greatest forecasting power for daily and monthly volatility, respectively, as it outperformed all other models including the GARCH model specifications. In this chapter we extend their research for daily stock market indices of six countries over the non-crash 1988 to 1999 period. We examine four different volatility models and find strong support for the SV model, although results appear to be sensitive to the choice of evaluation criteria.

Chapter 5 describes the concept of implied volatility, which is also often referred to as the market's perception of future volatility over the remaining life of the option. Implied volatility is obtained from option prices in conjuncture with a certain option pricing model and the accuracy of inferred implied volatility therefore not only depends on the efficiency with which the option market subsumes the available information but also on the use of the correct option pricing model, *i.e.* the model used by the market to price volatility. The interpretation of the well-known Black-Scholes option pricing model is discussed together with its underlying assumptions, one of which is that the volatility of the underlying asset is constant. The assumption of constant volatility would imply flat strike and term structures but substantial empirical research has shown that in practice this is not the case. Hence, the question arises which set of option series should be selected in order to obtain the market's volatility forecast with regard to the underlying asset. We give an overview of the literature in which the value of implied volatility as an information source for future volatility has been established empirically.

Chapter 6 compares the predictive ability of Stochastic Volatility (SV) models to that of volatility forecasts implied by option prices. A new variant of the SV model with implied volatility embedded in the variance equation is proposed and developed, which facilitates the use of statistical tests for nested models. The resulting model is referred to as the SVX model. The SVX model is then further extended to a volatility model with a persistence adjustment term; this is called the SVX^+ model. Provided that the market is efficient and that the option pricing model has been correctly specified, the information content of implied volatility should subsume that of all other variables in the information set. This hypothesis has been investigated in the context of GARCH models where findings were initially mixed, but more recent studies have indicated that implied volatility provides superior forecasts. We compare and contrast the performance of the various SV model specifications, both within sample and out-of-sample, for daily returns on the Standard & Poor's 100 index. The out-of-sample volatility forecasts are evaluated against daily squared returns as well as cumulative squared intraday returns for forecasting horizons ranging from 1 to 20 trading days.

The increasing availability of financial market data at intraday frequencies has not only led to the development of improved ex-post volatility measurements but has also inspired research into their value for the purpose of longer horizon volatility forecasts. Even though the theoretic underpinnings, developed by Andersen and Bollerslev (1998) and Barndorff-Nielsen and Shephard (2001), are involved, the calculation of intraday volatility measures is straightforward as squared intraday returns are cumulated to approximate realised volatility more closely. Theoretically, these intraday volatility measures should become free of measurement noise as the sampling frequency interval becomes infinitesimally small but the choice of sampling interval should be balanced against the effect of microstructure biases.

Chapter 7 explores the forecasting ability of models using high frequency data and the results are compared with a variety of volatility models for the Standard & Poor's 100 stock index. Two so-called realised volatility models, in which the intraday volatility process is modelled directly, are considered. The first is an unobserved components model, abbreviated to the UC-RV model, where actual volatility is modelled as an autoregressive moving average process. The second model is an autoregressive fractionally integrated moving average model, which allows for long memory in the logarithms of the intraday volatility measure; this is referred to as the ARFIMA-RV model. The predictive abilities of these two models are then compared with those of daily time-varying volatility models, such as the Stochastic Volatility (SV) model and the GARCH model, which are both extended to include the intraday volatility measure. Whereas thus far the emphasis has been on GARCH and ARFIMA-RV models we implement and empirically investigate a much wider range of models, both in sample and out-of-sample for different forecasting horizons and evaluation criteria. In addition, alternative intraday volatility measures are considered in order to account for the absence of intranight returns in stock markets.

Chapter 8 briefly summarises and provides concluding comments on the empirical findings presented in this book.

Chapter 2

ASSET RETURN VOLATILITY MODELS

Whereas the returns on financial assets can be assumed to be serially uncorrelated, although the issue of asset return predictability is still fiercely debated in the literature, see *e.g.* Campbell and MacKinlay (1999), it has become widely accepted that return volatility is highly autocorrelated. If the volatility process of financial asset returns were to be perfectly correlated, *i.e.* if volatility was constant over time, the variance of the underlying series could simply be calculated as

$$\sigma^2 = \frac{1}{T-1}\sum_{t=1}^{T}(y_t - m)^2, \tag{2.1}$$

where y_t is the continuously compounded return on the relevant asset at time t, T is the total number of observations and the sample mean $m = \frac{1}{T}\sum_{t=1}^{T} y_t$. A practical question that arises is what the length of the sample should be when volatility changes over time. Conventional wisdom would suggest that the sample should consist of a large number of observations as this would improve the statistical accuracy of the volatility estimates. On the other hand, it also implies that extreme shocks to the return process that took place a relatively long time ago, and which contain little information about the current volatility level, will still have a major impact since all observations in the sample are weighted equally. An alternative could be to select a smaller sample size and use a rolling window principle, where the sample size T remains constant and the variance is recalculated with each new observation that becomes available. The problem of extreme values remains however as the variance estimate of equation (2.1) will drop considerably in value when observations relating to this event fall out of the sample. Exponentially weighted methods have been proposed to remedy this problem but the reported high autocorrelation of the volatility process has led to the development of more intricate time-varying volatility models which incorporate this feature.

The remainder of this chapter is organised as follows. In the next section we discuss the salient features of stock index returns at different frequencies in more detail and in section 2.2 we give an overview of the two classes of time-varying volatility models which have received considerable attention in the literature. The models we discuss are the Generalised Autoregressive Conditional Heteroskedasticity (GARCH) and the Stochastic Volatility (SV) model which both describe the variance as an autoregressive structure. In section 2.3 the comparative abilities of GARCH and SV models are further investigated using empirical results of daily and weekly returns on the Financial Times All Share Index. The results obtained with the various volatility models are evaluated on the basis of their ability to capture the stylised facts of stock index return series as discussed in section 2.1, matching of theoretical moments, model fit and model assumptions.

2.1 Empirical Stylised Facts of Stock Index Return Series

The stock index return series we selected to work with throughout this section is the UK Financial Times All Shares Index over the period 2 January 1975 to 30 September 2000, yielding 6718 observations when we consider this series at a daily frequency. In addition, we examine the same series at weekly and monthly intervals, which results in samples that consist of 1344 and 310 observations respectively. For illustrative purposes we plot in figure 2.1 the continuously compounded returns on these three stock index series together with the accompanying histograms and we also present correlograms of the first 30 autocorrelation coefficients for the squared return series. The continuously compounded returns we express in percentage terms and these are calculated as

$$y_t = 100(\ln P_t - \ln P_{t-1}), \qquad t = 1, \ldots, T, \tag{2.2}$$

where P_t is the price of the underlying asset, *i.e.* the closing price of the stock market index, at time t. The graphs show that for all three time series the degree of dispersion around the mean value is not constant over time. Especially for the highest frequency series we are able to distinguish periods of prolonged low and high volatility, which is the phenomenon that is referred to in the literature as volatility clustering.

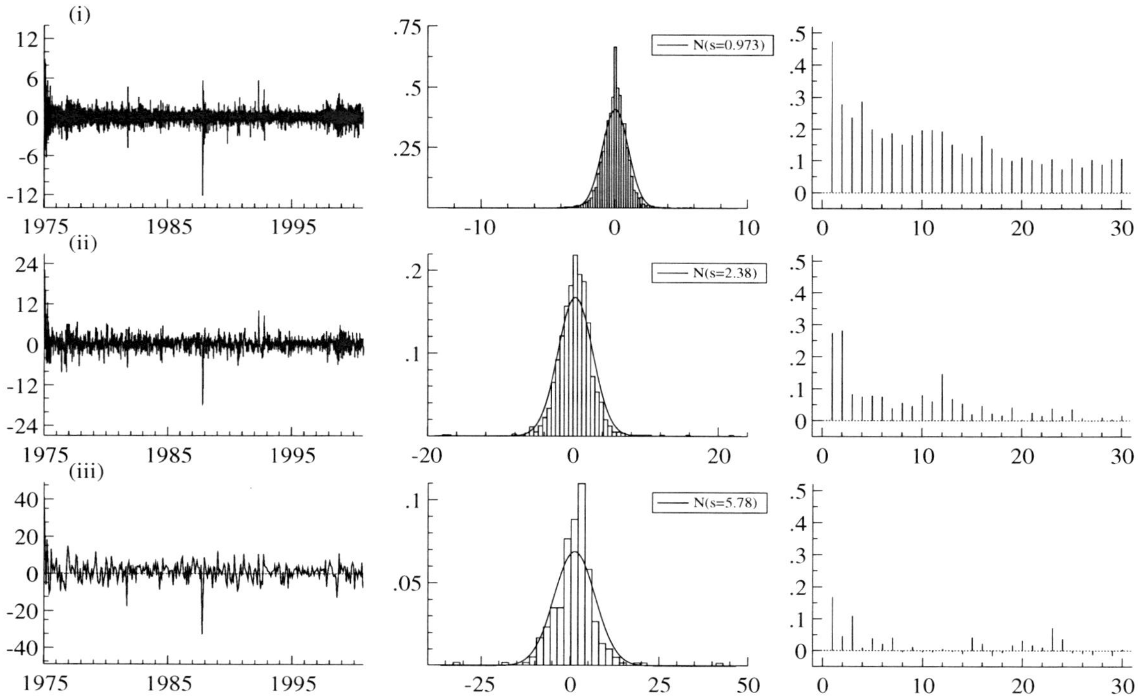

Figure 2.1. Returns for the Financial Times All Share Index (UK) at (i) daily, (ii) weekly and (iii) monthly frequencies between 02/01/75 and 30/09/00

Time series of returns are plotted on the left and histograms of returns, together with their normal approximation, are graphed in the centre. Graphs on the right are the correlograms of the first 30 autocorrelation coefficients of the squared return series.

Table 2.1. Summary statistics of returns y_t and squared returns y_t^2 on the FT All Share Index at daily, weekly and monthly frequencies between 2 January 1975 and 30 September 2000

Frequency No. of obs.	daily 6717		weekly 1343		monthly 309	
Series	y_t	y_t^2	y_t	y_t^2	y_t	y_t^2
Mean	0.057	0.951	0.284	5.760	1.234	34.894
Median	0.048	0.277	0.305	1.817	1.607	11.679
Variance	0.947	11.575	5.684	445.889	33.480	15517.09
Skewness	-0.192	20.874	0.318	14.883	0.477	10.993
Exc.Kurtosis	10.937	675.726	11.650	276.560	11.505	139.431
Maximum	8.943	146.828	22.090	487.971	42.133	1775.231
Minimum	-12.117	0.000	-17.453	0.000	-32.711	0.000
$\hat{\rho}_1$	0.159	0.472	0.102	0.274	0.086	0.168
$\hat{\rho}_2$	0.003	0.278	0.128	0.282	-0.113	0.047
$\hat{\rho}_3$	0.029	0.237	0.015	0.083	0.001	0.110
$\hat{\rho}_4$	0.039	0.287	0.027	0.075	0.015	0.009
$\hat{\rho}_5$	0.018	0.200	0.002	0.078	-0.077	0.039
$Q(12)$	252.02	4793.91	46.61	292.08	12.94	14.59
$Q(30)$	295.93	6341.83	75.09	314.95	32.05	18.52

$\hat{\rho}_\ell$ is the sample autocorrelation coefficient at lag ℓ with asymptotic standard error $1/\sqrt{T}$ and $Q(\ell)$ is the Box-Ljung portmanteau statistic based on ℓ squared autocorrelations. The critical values at the 1% significance level for the Q(12) and Q(30) statistic are 26.22 and 50.89, respectively.

The large negative value relating to the stock market crash of October 1987, when on a single trading day the stock index lost more than 12% of its value, is clearly observable for all three series. The histograms then suggest that y_t exhibits leptokurtosis and that the distribution of the returns is virtually symmetric around the mean, although the degree of asymmetry appears to increase as the sampling frequency decreases. In table 2.1 we report on the returns and the squared returns of the three series. We observe that the average daily return on the Financial Times All Share Index over the sample period amounts to 0.057% and that the average deviation from this value, calculated in terms of the variance as given in equation (2.1), is 0.947. These values can be translated into an average annual return of 14.8% and an annualised variation expressed in terms of standard deviations of 15.7%. The values for the third and fourth moment are in line with those typically found for high frequency financial asset return series, *i.e.* the series is negatively skewed and exhibits excess kurtosis. The skewness coefficient of -0.192 is relatively small but indicates that the distribution of the return series is asymmetric with more extreme observations situated left of the mean. The excess kurtosis value of 10.937 implies a leptokurtic distribution. Mandelbrot (1963) and Fama (1965) were the first to document that financial returns followed a leptokurtic distribution, which is characterised by fat tails and a narrow peak around the mean when compared with the normal distribution. This means that returns very close to the mean value are more likely, as are extreme changes in prices. It is especially this last feature that deserves special attention as these extreme values contribute considerably to increases in volatility.

As returns are expressed in terms of continuously compounded returns the multi-period returns are sums of the daily single period returns so average annualised returns for the weekly and monthly returns amount to the same 14.8% we observed for the daily returns. The median values then indicate that the weekly and monthly series have more observations to the right than to the left of their mean value. Unlike the daily return series, they have leptokurtic distributions that are positively, and not negatively, skewed.

The autocorrelation coefficients reported in table 2.1 show that the degree of autocorrelation is more pronounced for the squared returns than the returns themselves, although the monthly series have Q statistics that are relatively similar in value. In addition we observe that for the daily and weekly return series we are unable to reject the hypothesis that the first twelve and thirty autocorrelation coefficients are jointly equal to zero. Whereas highly autocorrelated returns reflect the effects of non-synchronous or thin trading, the high degree of autocorrelation found for the squared return series indicates that large (small) changes in returns are followed by large (small) changes of either sign. The correlograms in figure 2.1 show that the sample correlation coefficients of the daily squared returns are especially high and that their values decay very slowly as the lags increase. With the increase in sampling interval we observe that the degree of autocorrelation diminishes, so the higher the sampling frequency the more pronounced the degree of volatility clustering; the same can be deduced from the $Q(12)$ and $Q(30)$ statistics in table 2.1.

The stylised facts of these types of return series, which have also frequently been mentioned in the literature, can therefore be summarised as follows: the volatility of financial asset return series, such as stock index return series, is not constant over time and the returns experience prolonged periods of high and low volatility, often referred to as volatility clustering where volatility persistence is especially high for high frequency data, such as daily data. In addition, these series exhibit excess kurtosis when compared with the normal distribution. Although it has become accepted that the assumption that the returns themselves follow a random walk is a reasonable first approximation, it appears that the Financial Times All Share Index series might require that the returns themselves also be modelled.

2.2 Time-Varying Volatility Models

In the past two decades two prominent classes of models have been developed which capture the time-varying autocorrelated volatility process: the Generalised Autoregressive Heteroskedasticity (GARCH) and the Stochastic Volatility (SV) model. GARCH models define the time-varying variance as a deterministic function of past squared innovations and lagged conditional variances whereas the variance in the SV model is modelled as an unobserved component that follows some stochastic process.

Both the GARCH and the SV model are defined by their first and second moment, which are also referred to as the mean and the variance, respectively. The most general form of the mean equation can be written as

$$y_t = \mu_t + \sigma_t \varepsilon_t, \qquad \varepsilon_t \sim \mathrm{NID}(0,1), \quad t = 1, \ldots, T, \tag{2.3}$$

where y_t is the continuously compounded return on the relevant asset at time t and μ_t is its time-varying conditional mean. The disturbance term ε_t is independently and identically distributed with zero mean and unit variance. Usually, the assumption of a normal distribution for ε_t is added. The positive volatility process is then denoted by σ_t which remains to be specified in section 2.2.1 for GARCH and in section 2.2.2 for SV models. The mean adjusted series is therefore defined as white noise with unit variance multiplied by the volatility process σ_t. As the emphasis in this chapter will be on the volatility process we leave the mean unmodelled for compactness of exposition and redefine the mean equation as

$$y_t = \sigma_t \varepsilon_t, \qquad \varepsilon_t \sim \mathrm{NID}(0,1), \tag{2.4}$$

which is identical to equation (2.3) with μ_t equal to zero.

2.2.1 GARCH Models

Chronologically, the Autoregressive Conditional Heteroskedasticity model, introduced by Engle in his seminal 1982 paper about inflation in the United Kingdom, was the first model to allow for time-varying volatility. This model was subsequently generalised by Bollerslev (1986) who labelled it GARCH and after this numerous extensions followed[1]. The general form of the conditional variance of the GARCH(p, q) model, which has been used in countless empirical applications, is

$$\begin{aligned} \sigma_t^2 &= \omega + \sum_{i=1}^{q} \alpha_i y_{t-i}^2 + \sum_{i=1}^{p} \beta_i \sigma_{t-i}^2, \\ &= \omega + \sum_{i=1}^{q} \alpha_i (\sigma_{t-i}\varepsilon_{t-i})^2 + \sum_{i=1}^{p} \beta_i \sigma_{t-i}^2, \end{aligned} \tag{2.5}$$

which defines the variance as a linear function of the previous q squared innovations to the return process and p conditional variances. The parameters to be estimated are ω, $\alpha_1, \ldots, \alpha_q$ and $\beta_1, \ldots \beta_p$, whose values are restricted to $\omega > 0$, $\alpha_i \geq 0$ and $\beta_i \geq 0$[2]. Engle's original ARCH(q) model can then be obtained by setting p equal to zero[3]. GARCH models are conditional models in the sense that the variance is calculated subject to an information set I_{t-1} which consists of return observations prior to time t. In the literature GARCH models have therefore frequently been referred to as observation-driven models.

The most commonly used model in applied financial studies is the GARCH(1, 1) model which is given by

$$\sigma_t^2 = \omega + \alpha(\sigma_{t-1}\varepsilon_{t-1})^2 + \beta\sigma_{t-1}^2, \tag{2.6}$$

---

[1]We refer to these models collectively as GARCH models. For surveys on the extensive GARCH literature we refer to Bollerslev, Chou and Kroner (1992), Bera and Higgins (1993), Bollerslev, Engle and Nelson (1994), Diebold and Lopez (1995) and to the comprehensive selection of influential (G)ARCH papers in Engle (1995).

[2]In order to ensure that the process is covariance stationary $\sum_{i=1}^{q} \alpha_i + \sum_{j=1}^{p} \beta_j$ is set to be less than one. The positivity constraints have been questioned by Nelson and Cao (1992) who give necessary and sufficient conditions to ensure a non-negative variance which are less stringent. These weaker conditions are however mainly relevant for the higher order GARCH models.

[3]The development of the GARCH model was primarily motivated by the fact that in empirical applications high orders of q were needed.

with parameter values restricted as above. Provided that the sum of α and β is less than one, the unconditional expectation of the conditional variance is constant and finite and given by

$$\begin{aligned} \mathrm{E}(\sigma_t^2) &= \omega + \alpha\,\mathrm{E}(y_{t-1}^2) + \beta\,\mathrm{E}(\sigma_{t-1}^2) \\ &= \omega + (\alpha+\beta)\,\mathrm{E}(\sigma_{t-1}^2) \\ &= \frac{\omega}{1-\alpha-\beta}, \end{aligned} \tag{2.7}$$

using $\mathrm{E}(y_t^2) = \mathrm{E}(\sigma_t^2)$ and the law of iterated expectations. In empirical financial research with high frequency data $(\alpha+\beta)$ is often estimated as being close to unity which implies a high degree of volatility persistence. Apart from volatility clustering, GARCH models also capture part of the excess kurtosis observed in financial time series. Under the assumption of normality, existence of the fourth order moment for the GARCH(1, 1) model is ensured if $\beta^2 + 2\alpha\beta + 3\alpha^2 < 1$. Subject to this restriction it can be shown that the fourth moment will exhibit excess kurtosis as

$$\begin{aligned} \kappa_y &= \frac{\kappa_\varepsilon\,\mathrm{E}(\sigma_t^4)}{\mathrm{E}(\sigma_t^2)^2} \\ &= \frac{3\omega^2(1+\alpha+\beta)[(1-\alpha-\beta)(1-\beta^2-2\alpha\beta-3\alpha^2)]^{-1}}{\omega^2[(1-\alpha-\beta)^2]^{-1}} \\ &= \frac{3(1-\beta^2-2\alpha\beta-\alpha^2)}{(1-\beta^2-2\alpha\beta-3\alpha^2)} \\ &= 3 + \frac{6\alpha^2}{1-\beta^2-2\alpha\beta-3\alpha^2}, \end{aligned} \tag{2.8}$$

and therefore $\kappa_y > \kappa_\varepsilon$; see Bollerslev (1986) and He and Teräsvirta (1999).

Equation (2.6) can also be represented by

$$\begin{aligned} \sigma_t^2 &= \omega + (\alpha+\beta)\sigma_{t-1}^2 + \alpha(y_{t-1}^2 - \sigma_{t-1}^2) \\ &= \omega + (\alpha+\beta)\sigma_{t-1}^2 + \alpha(\sigma_{t-1}^2\varepsilon_{t-1}^2 - \sigma_{t-1}^2), \end{aligned} \tag{2.9}$$

which clearly shows that volatility persistence in the GARCH(1, 1) model is measured by $(\alpha+\beta)$. The $(\sigma_{t-1}^2\varepsilon_{t-1}^2 - \sigma_{t-1}^2)$ term can be thought of as the shock to the volatility process where the α parameter measures the extent to which this period's volatility is influenced by last period's shock to the returns process. An unanticipated shock to the return process at time $t-1$ is therefore not incorporated into the volatility process until time t.

When we write equation (2.6) in terms of squared returns we obtain

$$\begin{aligned} y_t^2 &= \omega + (\alpha+\beta)y_{t-1}^2 - \beta(y_{t-1}^2 - \sigma_{t-1}^2) + (y_t^2 - \sigma_t^2) \\ &= \omega + (\alpha+\beta)\sigma_{t-1}^2\varepsilon_{t-1}^2 - \beta(\sigma_{t-1}^2\varepsilon_{t-1}^2 - \sigma_{t-1}^2) + \\ &\quad (\sigma_t^2\varepsilon_t^2 - \sigma_t^2), \end{aligned} \tag{2.10}$$

which reveals the dynamics of the GARCH$(1,1)$ model as that of an ARMA$(1,1)$ process. Extending this GARCH model to include further lags results in the GARCH(p,q) model as given in equation (2.5) where p and $\max(p,q)$ are the order of the moving average and autoregressive component of the GARCH process, respectively. It is also evident from the above notation that Engle's ARCH(q) model follows an AR(q) process. Given that y_t^2 follows an ARMA$(1,1)$ process for the GARCH$(1,1)$ model with $(\alpha+\beta)$ the AR(1) and β the MA(1) parameter, its autocorrelation function can be evaluated accordingly and therefore

$$corr(y_t^2, y_{t-1}^2) = \rho_1 = \frac{\alpha(1-\alpha\beta-\beta^2)}{1-2\alpha\beta-\beta^2} \tag{2.11}$$

and

$$corr(y_t^2, y_{t-s}^2) = \rho_s = \rho_1(\alpha+\beta)^{s-1} \qquad s > 1. \tag{2.12}$$

Equation (2.9) shows that the variance of the GARCH model reacts symmetrically to shocks to the return process as these are squared, therefore only taking into account the magnitude of the shock and not its sign. Research, dating back to Black (1976), has shown however that unexpected stock returns are negatively correlated with volatility, *i.e.* volatility decreases when returns are higher then expected whereas bad news leads to increases in volatility. The EGARCH model proposed by Nelson (1991) allows for this leverage effect, or asymmetric volatility. The return volatility for the EGARCH$(1,1)$ model is defined as

$$\begin{aligned} \ln\sigma_t^2 &= \omega + \alpha(|\varepsilon_{t-1}| - \mathrm{E}\,|\varepsilon_{t-1}|) + \gamma\varepsilon_{t-1} + \beta\ln\sigma_{t-1}^2 \\ &= \omega + \alpha(|\varepsilon_{t-1}| - \sqrt{\frac{2}{\pi}}) + \gamma\varepsilon_{t-1} + \beta\ln\sigma_{t-1}^2. \end{aligned} \tag{2.13}$$

Specification of the EGARCH model in logarithmic form adds to the attractiveness of the model as the parameters no longer need to be restricted to positive values to ensure a non-negative conditional variance. The $\alpha(|\varepsilon_{t-1}| - \sqrt{\frac{2}{\pi}})$ term captures the magnitude effect as found in the GARCH model with α assumed to be positive, in which case $\ln \sigma_t^2$ increases (decreases) when $|\varepsilon_{t-1}|$ is larger (smaller) than the expected value. The leverage effect is accounted for by $\gamma\varepsilon_{t-1}$, provided that the asymmetry parameter is negative. If $\gamma < 0$, then a negative standardised residual, *i.e.* return, at time $t-1$ will lead to an increase in the conditional variance, whereas a positive return will result in a decreasing conditional variance.

The most common procedure used for estimating the parameters in GARCH models involves the maximisation of a likelihood function, the so-called maximum likelihood (ML) method, where the function has to be maximised with regard to all parameters in the model. Above we mentioned the ability of GARCH models to capture excess kurtosis when the standardised residual is assumed to follow a normal distribution. Unfortunately the model does not fully compensate for the degree of leptokurtosis found in financial time series. It is however possible to adopt alternative error distributions which display fatter tails than the normal distribution such as the Student-t distribution proposed by Bollerslev (1987)[4].

2.2.2 SV Models

The second class of volatility model which will be discussed in this chapter is the so-called Stochastic Volatility (SV) model[5]. The most popular version of the SV model defines volatility as a logarithmic first-order autoregressive process, which is a discrete-time approximation of the continuous-time Ornstein-Uhlenbeck diffusion process used in the option pricing literature[6]. The variance equation of the SV model is then specified as

$$\sigma_t^2 = \sigma^{*2} \exp(h_t) \tag{2.14}$$

[4]Nelson (1991) suggested that ε_t follow a Generalised Error Distribution (GED) which is a family of distributions including the normal distribution for $v = 2$ and fatter tailed distributions for $v < 2$.

[5]SV models are reviewed in, for example, Taylor (1994), Ghysels, Harvey and Renault (1996) and Shephard (1996).

[6]See Hull and White (1987), Scott (1987), Wiggins (1987) and Chesney and Scott (1989).

with positive scaling factor σ^{*2}. It follows that $h_t = \ln(\sigma_t^2/\sigma^{*2})$ where the stochastic process for h_t is

$$h_t = \phi h_{t-1} + \sigma_\eta \eta_t \qquad \eta_t \sim \mathrm{NID}(0,1), \tag{2.15}$$

with persistence parameter ϕ which is restricted to a positive value less than one to ensure stationarity. The disturbances ε_t and η_t are mutually uncorrelated, contemporaneously and at all lags[7]. The unconditional variance implied by the SV model is given by

$$\begin{aligned} \mathrm{E}(\sigma_t^2) &= \sigma^{*2}\exp(\mathrm{E}(h_t) + 0.5\,\mathrm{V}(h_t)) \\ &= \sigma^{*2}\exp\left(0.5\frac{\sigma_\eta^2}{1-\phi^2}\right), \end{aligned} \tag{2.16}$$

with unconditional mean $\mathrm{E}(h_t)$ and variance $\mathrm{V}(h_t)$ equal to zero and $\sigma_\eta^2/(1-\phi^2)$, respectively[8]. It can be shown that the SV model also captures part of the excess kurtosis as

$$\begin{aligned} \kappa_y &= \frac{\kappa_\varepsilon\,\mathrm{E}(\sigma_t^4)}{\mathrm{E}(\sigma_t^2)^2} \\ &= \frac{3\sigma^{*4}\exp(2\,\mathrm{V}(h_t))}{\sigma^{*4}\exp(\mathrm{V}(h_t))} \\ &= 3\exp(\mathrm{V}(h_t)) \\ &= 3\exp\left(\frac{\sigma_\eta^2}{1-\phi^2}\right), \end{aligned} \tag{2.17}$$

which also implies that $\kappa_y > \kappa_\varepsilon$ given that $\exp(\mathrm{V}(h_t)) > 1$. Of additional interest is the correlation of the squared returns implied by the SV model at different lag lengths since significant autocorrelation coefficients for y_t^2 are often observed in financial time series, even at quite extended lags. The covariance of the squared observations can be written as

$$cov(y_t^2 y_{t-s}^2) \;=\; \mathrm{E}(y_t^2 y_{t-s}^2) - \mathrm{E}(\mathrm{E}(y_t^2)\,\mathrm{E}(y_{t-s}^2))$$

---

[7]If the correlation coefficient between ε_t and η_{t+1} is estimated and found to be negative, a leverage effect similar to that discussed with regard to the EGARCH model is observed. Also see the discussion and references in Harvey and Shephard (1996).

[8]Derivation of the moments of σ_t becomes rather straightforward given that $\mathrm{E}(\sigma_t^r) = \sigma^{*r}\exp(0.5r\,\mathrm{E}(h_t) + 0.125r^2\,\mathrm{V}(h_t))$ for all positive integer values r.

$$\begin{aligned} &= \sigma^{*4}\exp((1+\phi^s)\,\mathrm{V}(h_t)) - \sigma^{*4}\exp(\mathrm{V}(h_t)) \\ &= \left[\sigma^{*4}\exp(\mathrm{V}(h_t))\right]\left[\exp(\phi^s\,\mathrm{V}(h_t)) - 1\right], \end{aligned}$$

and the variance of the squared observations is

$$\begin{aligned} \mathrm{V}(y_t^2) &= 3\sigma^{*4}\exp(2\,\mathrm{V}(h_t)) - \sigma^{*4}\exp(\mathrm{V}(h_t)) \\ &= \left[\sigma^{*4}\exp(\mathrm{V}(h_t))\right]\left[3\exp(\mathrm{V}(h_t)) - 1\right]. \end{aligned}$$

Therefore the autocorrelation coefficient at lag s takes the following form

$$\begin{aligned} corr(y_t^2, y_{t-s}^2) &= \rho_s = \frac{\left[\sigma^{*4}\exp(\mathrm{V}(h_t))\right]\left[\exp(\phi^s\,\mathrm{V}(h_t)) - 1\right]}{\left[\sigma^{*4}\exp(\mathrm{V}(h_t))\right]\left[3\exp(\mathrm{V}(h_t)) - 1\right]} \\ &= \frac{\exp(\phi^s\,\mathrm{V}(h_t)) - 1}{3\exp(\mathrm{V}(h_t)) - 1} \\ &\approx \phi^s \left(\frac{\exp\left(\frac{\sigma_\eta^2}{1-\phi^2}\right) - 1}{3\exp\left(\frac{\sigma_\eta^2}{1-\phi^2}\right) - 1} \right), \end{aligned} \qquad (2.18)$$

where the above approximation holds when ϕ^s is close to unity and/or $\mathrm{V}(h_t)$ is very close to zero; see Taylor (1986, 1994).

Alternative specifications for the SV model can be deduced from

$$\begin{aligned} \ln\sigma_t^2 &= \ln\sigma^{*2} + h_t \\ &= \ln\sigma^{*2} + \phi(\ln\sigma_{t-1}^2 - \ln\sigma^{*2}) + \sigma_\eta\eta_t \\ &= (1-\phi)\ln\sigma^{*2} + \phi\ln\sigma_{t-1}^2 + \sigma_\eta\eta_t. \end{aligned}$$

The main distinction between GARCH and SV models is that the latter has separate disturbance terms in the mean and variance equation, ε_t and η_t, respectively, which precludes direct observation of the variance process σ_t^2. GARCH models are deterministic in the sense that only the mean equation has a disturbance term and that its variance is modelled conditionally on I_{t-1}, that is the information up to and including time $t-1$. Therefore, the variance can be observed at time t. For the SV model, the deviation of y_t from the mean is captured by a function of the two disturbance terms whereas in the GARCH model this deviation is accounted for by a single disturbance term. For the GARCH model this point is evident but to clarify this for the SV model, we rewrite the model as follows:

$$\begin{aligned} y_t &= \sigma_t \varepsilon_t \\ &= \sigma^* \exp(0.5h_t)\varepsilon_t \\ &= \sigma^* \exp(0.5\phi h_{t-1}) \exp(0.5\eta_t)\varepsilon_t. \end{aligned}$$

The overall innovation term of the SV model is the error term $\exp(0.5\eta_t)\varepsilon_t$ with a zero mean but with a non-Gaussian density.

Although SV models are seen as a competitive alternative to GARCH models their empirical application has been limited. This can mainly be attributed to the difficulties that arise as a result of the intractability of the likelihood function which prohibits its direct evaluation. However, in recent years considerable advances have been made in this area. The estimation techniques that have been proposed for SV models can be divided into two groups: those that seek to construct the full likelihood function and those that approximate it or avoid the issue altogether. The methods originally suggested by Taylor (1986) and Harvey, Ruiz and Shephard (1994) belong to the latter category. Recently attention has moved towards the development of techniques that attempt to evaluate the full likelihood function[9]. For recent reviews on these full likelihood methods we refer to Fridman and Harris (1998), Sandmann and Koopman (1998) and Pitt and Shephard (1999), amongst others.

2.3 Empirical Applications of Time-Varying Volatility Models

GARCH models and their various extensions have been widely applied to financial and economic time series and the empirical GARCH literature has grown dramatically over the past two decades. Initially many of these stock return studies investigated the lag lengths p and q in the GARCH(p, q) model and concluded that for most empirical implementations it was sufficient to adopt low orders[10]. Ever since the GARCH(1,1) model has become the standard GARCH model specification. The predominance of the first-order autoregressive SV model on the other hand is probably due to the fact that even this model presents

[9]See, for example, Jacquier, Polson and Rossi (1994), Kim, Shephard and Chib (1998), Sandmann and Koopman (1998) and Fridman and Harris (1998).

[10]See *e.g.* French, Schwert and Stambaugh (1987) and Akgiray (1989) who examined the US stock market and Poon and Taylor (1992) for results with regard to the UK stock market.

a challenge in terms of estimation[11]. In this section we fit these GARCH and SV model specifications to the daily and weekly Financial Times All Share Index return series introduced in section 2.1. We do not estimate the models at the monthly frequency as volatility persistence effects are known to weaken when the data is sampled less frequently[12]. Our objective is to investigate the comparative abilities of the two classes of volatility models in capturing the stylised facts of stock index returns such as leptokurtosis and volatility clustering as well as to test the underlying model assumptions. The mean equation of both volatility models is then defined as

$$y_t = \sigma_t \varepsilon_t, \qquad \varepsilon_t \sim \text{NID}(0,1), \quad t = 1, \ldots, T,$$

and volatility equations are given by

$$\textit{GARCH(1,1)}: \qquad \sigma_t^2 = \omega + \alpha y_{t-1}^2 + \beta \sigma_{t-1}^2$$

$$\textit{SV:} \qquad \ln \sigma_t^2 = (1-\phi)\ln \sigma^{*2} + \phi \ln \sigma_{t-1}^2 + \sigma_\eta \eta_t,$$

with parameter restrictions and distributional assumptions as discussed in section 2.2. As research in the GARCH literature has indicated that the assumption of normally distributed error terms is often empirically unjustified the GARCH(1,1) model is also estimated with ε_t following a Student-t instead of a Gaussian distribution. This problem is far less pronounced for SV models due to their greater flexibility in coping with outliers[13].

The results for the daily return series are given in table 2.2 where we present the parameter estimates and their corresponding 95% confidence intervals together with some diagnostics[14]. After the models have been

---

[11]For the estimation of the SV model we use the exact maximum likelihood methods as described in Appendices A and B. Computing times for estimating SV models is usually higher than for estimating GARCH models which is due to the relatively more complicated methods required for estimating SV models.

[12]See for example Baillie and Bollerslev (1989). For our data set this is confirmed by the insignificant $Q(12)$ and $Q(30)$ statistics for the monthly squared returns in table 2.1.

[13]For recent research on SV models with fat-tailed error distributions we refer to Liesenfeld and Jung (2000) and Jacquier, Polson and Rossi (2001).

[14]The data series together with the program used for the estimation of the SV models can be found on the Internet at `www.ssfpack.com/holbook`; also see Appendix C.

Table 2.2. Estimation results for daily returns on the Financial Times All Share Index

Model No. of obs. T	SV Model 6717		GARCH Model 6717	
σ^{*2}	0.653 0.525 0.812	ω	0.018 0.012 0.024	0.013 0.008 0.018
ϕ	0.986 0.977 0.991	α	0.086 0.071 0.102	0.077 0.062 0.092
σ_η^2	0.016 0.011 0.023	β	0.893 0.874 0.913	0.907 0.889 0.926
		υ		11.634 9.123 14.146
		$\alpha+\beta$	0.980	0.984
$\ln L$	$-8,267.21$		$-8,367.67$	$-8,255.13$
$LR(\upsilon=0)$				225.08
AIC	16,540.43		16,741.34	16,518.26
$Q(12)$	150.58		202.80	203.78
$Q(30)$	171.26		220.90	221.74
N	22.2		7,448.1	8,212.7
σ_∞^2	0.875		0.866	0.826
κ_y	5.402		4.747	5.623

Parameter estimates are reported together with the asymptotic 95% confidence intervals which are asymmetric for σ^{*2}, ϕ and σ_η^2; υ is the degrees of freedom for the conditional Student t-distribution. $LR(\upsilon = 0)$ is the likelihood ratio statistic for the hypothesis $\upsilon = 0$; AIC is the Akaike Information Criterion which is calculated as -2(ln L) + $2p$; $Q(\ell)$ is the Box-Ljung portmanteau statistic for the estimated observation errors which is asymptotically χ^2 distributed with $\ell - p$ degrees of freedom where p is the total number of estimated parameters; N is the χ^2 normality test statistic with 2 degrees of freedom. σ_∞^2 and κ_y denote the unconditional variance and the fourth moment as implied by the volatility models.

fitted the standardised residuals ε_t are obtained and these serve as input for the Box-Ljung $Q(\ell)$ and the N statistics. With the $Q(\ell)$ statistic we test for remaining autocorrelation in ε_t at lag ℓ whereas with the χ^2 normality statistic we examine to what extent the normality assumption with regard to ε_t is abided. The theoretical unconditional variance and the kurtosis values are calculated from equations (2.16) and (2.17) for the SV model and from equations (2.7) and (2.8) for the GARCH models where in the case of the t-distributed GARCH(1,1) model we have $\kappa_\varepsilon = 3(\hat{\upsilon} - 2)(\hat{\upsilon} - 4)^{-1}$, see Bollerslev (1987).

The table shows that all parameter estimates in both the SV and GARCH models are statistically significant. In the SV model stationarity is ensured when $\phi < 0$ which we imposed and we observe that the estimate for ϕ is also significantly different from unity. As $\hat{\phi} = 0.986$ is very close to one, the SV model indicates a high degree of volatility persistence for the daily returns on the Financial Times All Share Index. This confirms findings in the SV literature by *e.g.* Jacquier *et al.* (1994), Sandmann and Koopman (1998) and Fridman and Harris (1998) who report on persistence estimates typically ranging from 0.96 to 0.99 for daily Standard & Poor's 500 returns. GARCH(1,1) models, hereafter simply referred to as GARCH models, are stationary when the sum of α and β is smaller than one. In our analysis this is always the case as $(\hat{\alpha} + \hat{\beta})$ is equal to 0.980 in the GARCH model with conditional normally distributed error terms and equal to 0.984 in its Student-t distributed counterpart. The parameter estimate for υ is then found to be statistically significant as is the likelihood ratio test statistic testing for $\upsilon = 0$; this suggests that a better GARCH model fit is obtained when ε_t is assumed to follow a Student t-distribution.

Since SV and GARCH models are not nested we cannot perform LR tests to distinguish between the two classes of models. The Akaike Information Criterion (AIC) can be used to evaluate the models. We find that the Gaussian GARCH model is outperformed by the SV model but that the GARCH model with t-distributed error terms has the best AIC value. Danielsson (1994) was one of the first to compare the model fit of a number of SV model specifications with those of various GARCH models. On the basis of Schwarz criteria he concluded that the so-called dynamic SV model, which effectively is the logarithmic first-order autoregressive SV model, outperformed all GARCH models[15]. He did however not use t-distributed GARCH models. Subsequent studies such as the one by Kim, Shephard and Chib (1998) did. They reported that the Gaussian GARCH model did not fit their exchange rate data particularly well and that the fitted log-likelihood of the t-distributed GARCH model was slightly higher than that of the SV model, although at the cost of an additional model parameter. Shephard (1996) also conducted a comparative study which included the t-distributed GARCH model and showed that this was the preferred volatility model immediately fol-

[15] The Schwarz Criterion (SC) $= -2(\ln L) + (\ln T)(p)$ where T denotes the sample size and p the number of parameters in the model. The SC is comparable with the AIC which has second term $(2p)$ instead of $(\ln T)(p)$.

lowed by the SV model specification which had similar diagnostics. His findings with regard to stock index data are partially reflected here. We observe the best Box-Ljung Q statistics for the SV model which indicates that the assumption of zero autocorrelation of ε_t is best abided by the SV model[16]. Although the normality statistics always exceed their critical value and the hypothesis of a normally distributed standardised error term has to be rejected for all models, the SV model has values for N that are substantially lower than those of the GARCH models. Further, in section 2.2 it was shown that both SV and GARCH models were capable of absorbing some of the excess kurtosis found in the return series which is confirmed by the values for κ_y reported here as these are larger than 3 for all models. The SV and t-distributed GARCH model then capture most of the excess kurtosis; therefore these models are better-suited for accommodating large outliers in the data series than Gaussian GARCH models[17].

Estimation results for the weekly Financial Times All Share Index series are presented in table 2.3. For both the SV and the GARCH models the degree of volatility persistence has decreased which is to be expected as the data series is less frequently sampled. Many of the observations we made with regard to the daily data also apply here to the weekly series: both classes of volatility model capture the well-documented phenomenon of volatility clustering and the AIC value is again lowest for the t-distributed GARCH model, although the difference with the SV model has become smaller due to the reduced sample size. Distributional model assumptions for ε_t are again least violated by the SV model and the unconditional expectation of the fourth moment denoted by κ_y is again closest in conformance with the raw data for the SV and the t-distributed GARCH model.

Following Jacquier *et al.* (1994) we plot in figure 2.2 the theoretical autocorrelations coefficients of the fitted Gaussian SV and GARCH models together with the sample autocorrelation coefficients of the squared return series. We do this in order to investigate the ability of both classes of volatility models to account for the autocorrelation structure in the squared returns. In the case of the GARCH model the theoretical

[16] As $\hat{\rho_1} = 0.159$ in table 2.1 we expect that modelling the mean as a first-order autoregressive process might lead to further reduction in the values for $Q(\ell)$.

[17] We note that the kurtosis value which we report is the theoretical value implied by the volatility model parameters, which we want to be as close as possible to the sample kurtosis of returns, and not the frequently reported sample kurtosis of ε_t which under the normality assumption should ideally be centred around 3. It is evident from the encompassing N statistics that the sample kurtosis of the standardised error term in the SV model most closely approaches this value.

Table 2.3. Estimation results for weekly returns on the Financial Times All Share Index

Model No. of obs. T	SV Model 1343		GARCH Model 1343	
σ^{*2}	4.147	ω	0.296	0.311
	3.041 5.655		0.117 0.476	0.111 0.512
ϕ	0.966	α	0.105	0.107
	0.929 0.984		0.072 0.139	0.059 0.154
σ_η^2	0.037	β	0.838	0.828
	0.020 0.066		0.784 0.893	0.756 0.900
		υ		8.392
				5.452 11.332
		$\alpha + \beta$	0.944	0.935
$\ln L$	$-2,900.94$		$-2,940.76$	$-2,896.45$
$LR(\upsilon = 0)$				88.62
AIC	5,807.88		5,887.51	5,800.91
$Q(12)$	16.36		25.44	24.90
$Q(30)$	37.73		48.02	48.02
N	6.849		2,161.9	2,237.0
σ_∞^2	5.447		5.265	4.807
κ_y	5.175		3.766	5.036

Parameter estimates are reported together with the asymptotic 95% confidence intervals which are asymmetric for σ^{*2}, ϕ and σ_η^2. $LR(\upsilon = 0)$ is the likelihood ratio statistic for the hypothesis $\upsilon = 0$; AIC is the Akaike Information Criterion which is calculated as -2(ln L) + $2p$; $Q(\ell)$ is the Box-Ljung portmanteau statistic for the estimated observation errors which is asymptotically χ^2 distributed with $\ell - p$ degrees of freedom where p is the total number of estimated parameters; N is the χ^2 normality test statistic with 2 degrees of freedom. σ_∞^2 and κ_y denote the unconditional variance and the fourth moment as implied by the volatility models.

autocorrelation function (ACF) decays at an exponential rate of $(\hat{\alpha} + \hat{\beta})$ whereas for the SV model the exponential rate of decay in the ACF is determined by the ϕ parameter estimate. Figure 2.2 shows that the ACF of the GARCH models decays more rapidly than that of the SV models both at the daily and weekly frequency which is a direct consequence of the fact that $(\hat{\alpha} + \hat{\beta})$ is always smaller than $\hat{\phi}$ in tables 2.2 and 2.3.

Unlike Jacquier *et al.* (1994), who constructed a similar plot for weekly returns on a portfolio formed of the smallest decile of listed New

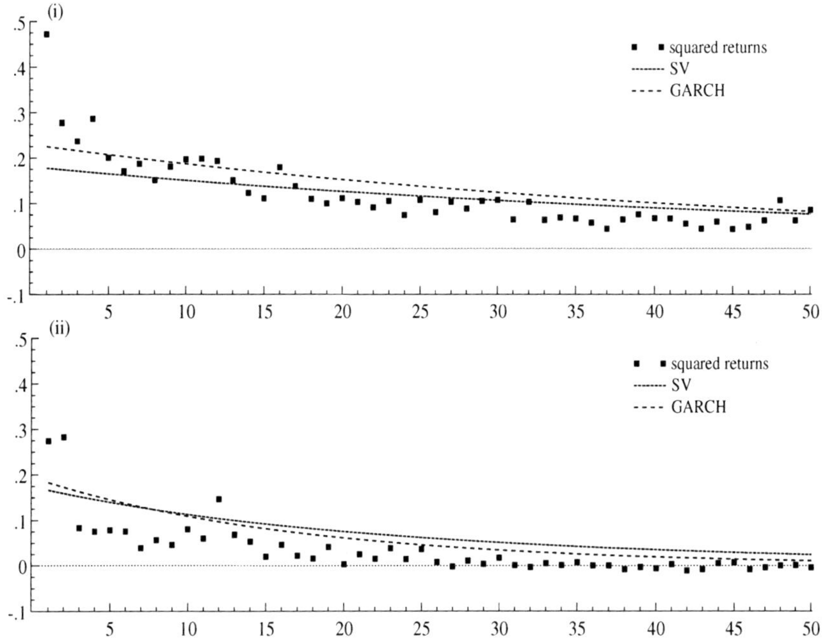

Figure 2.2. Autocorrelation coefficients for the squared returns, SV and GARCH models of the Financial Times All Share Index (UK) at (i) daily and (ii) weekly frequencies
Autocorrelation coefficients of the squared return series y_t^2 are calculated for the full Financial Times All Share Index series between 02/01/75 and 30/09/00; the implied theoretical autocorrelations of the fitted SV and GARCH model are calculated from the parameter estimates in tables 2.2 and 2.3 using equations (2.18) for the SV model and equations (2.11) and (2.12) for the GARCH model.

York Stock Exchange stocks over the period 1962 to 1991, we cannot conclude that the implied ACF of the SV model is in closer conformance with the ACF of the squared returns either at the daily or the weekly frequency[18]. Instead we observe that both the SV and the GARCH model appear to follow the actual data quite reasonably, although for both models the theoretical values for ACF(ℓ) are not high enough for small values of ℓ and remain too high when the lag length increases. We

[18]We note that their SV model estimate of the volatility persistence term is equal to 0.91 which explains the rapid exponential decay of the ACF (see figure 1); the high implied first-order autocorrelation coefficient for the SV model is then mainly attributable to the relatively high estimate for the variance of the log volatility process. Also see the discussion in Ghysels *et al.* (1996).

expect this problem to become more severe if we were to assume that ε_t followed a Student-t distribution. The relationship between the theoretical first-order autocorrelation coefficient and the theoretical kurtosis is investigated in the SV model context by Liesenfeld and Jung (2000). They show that the theoretical ACF can be expressed as a function of theoretical kurtosis, where

$$\rho_s = \frac{(\kappa_y/\kappa_\varepsilon)^{\phi^s} - 1}{\kappa_y - 1},$$

with $\kappa_\varepsilon = 3(\hat{\upsilon}-2)(\hat{\upsilon}-4)^{-1}$. If at the daily frequency we assumed, ceteris paribus, $\hat{\upsilon} = 12$, then $\kappa_y = 6.73$ and the theoretical ACF(1) of the t-distributed SV model is equal to 0.136 which is smaller than the 0.178 for the Gaussian SV model[19]. The same applies to the GARCH models where assuming a Student-t distribution for ε_t increases the value for κ_y ceteris paribus which in turn ceteris paribus decreases the value for the ACF(1) of the t-distributed GARCH model when compared to the Gaussian GARCH model.

In summary, our findings in tables 2.2 and 2.3 suggest that the Gaussian GARCH model is the least successful volatility model. Of the two remaining models the t-distributed GARCH model appears to give a slightly better model fit than the SV model but does so at the cost of an additional model parameter. Diagnostics for ε_t are much better in the SV model and values for κ_y are comparable. Hence we feel that both the SV and the t-distributed GARCH model fit the data equally well when evaluated on the basis of the results presented in tables 2.2 and 2.3. However, when we consider the autocorrelation of the squared returns and compare its values with those of the theoretical autocorrelation functions of the Gaussian SV and GARCH models we observe that both models follow the sample autocorrelations reasonably well but that their performance deteriorates once leptokurtic distributions are imposed. Taking all the above criteria into account this then suggests that the SV model best captures the salient features of the stock index return volatility.

[19] Also see the discussion in Ghysels *et al.* (1996) who show that ρ_s "is at maximum for a normal distribution".

Chapter 3

THE STOCHASTIC VOLATILITY IN MEAN MODEL: EMPIRICAL EVIDENCE FROM INTERNATIONAL STOCK MARKETS

3.1 Introduction

The Stochastic Volatility (SV) models we present in this chapter are a practical alternative to the Generalised Autoregressive Conditional Heteroskedasticity (GARCH) type models that have been used so widely in empirical financial research and which have relied on simultaneous modelling of the first and second moment[1]. For certain financial time series such as stock index returns, which have been shown to display high positive first-order autocorrelations, this constitutes an improvement in terms of efficiency; see Campbell, Lo and MacKinlay (1997, Chapter 2). The volatility of daily stock index returns has been estimated with SV models but usually results have relied on extensive pre-modelling of these series, thus avoiding the problem of simultaneous estimation of the mean and variance[2]. New estimation techniques now enable us to include explanatory variables in the mean equation and estimate their coefficients simultaneously with the parameters of the volatility process[3]. One of the explanatory variables in our model is the variance process itself, hence its name: Stochastic Volatility in Mean (SVM).

---

[1]The material in this chapter was previously published in the *Journal of Applied Econometrics* (2002) **17**, 667-689, as "The Stochastic Volatility in Mean Model: Empirical evidence from international stock markets" by S.J. Koopman and E. Hol Uspensky. Reprinted with permission of John Wiley & Sons Limited.

[2]The same seasonally adjusted S&P Composite stock index series (Gallant, Rossi and Tauchen, 1992) has been used in a number of studies, see for example: Jacquier, Polson and Rossi (1994), Danielsson (1994), Sandmann and Koopman (1998), Fridman and Harris (1998) and Chib, Nardari and Shephard (1998).

[3]Also see Fridman and Harris (1998) and Chib *et al* (1998).

The development of an SV model that includes volatility as one of the determinants of the mean makes our model suitable for empirical applications in which returns are partially dependent on volatility, such as studies that investigate the relationship between the mean and variance of stock returns. The SVM model can therefore be viewed as the SV counterpart of the ARCH-M model of Engle, Lilien and Robins (1987). The main difference between the two classes of models is that the ARCH-M model intends to estimate the relationship between expected returns and expected volatility, whereas the aim of the SVM model is the simultaneous estimation of the *ex ante* relation between returns and volatility and the volatility feedback effect. This is further discussed in section 3.3. We investigate the intertemporal relationship between excess returns on a daily stock index and their volatility for three international stock indices and compare the resulting parameter estimates of the SVM model with those obtained for the GARCH-M model. The stock indices we examine are the Financial Times All Share (UK), the Standard & Poor's Composite (US) and the Topix Index (Japan).

The remainder of this chapter is organised as follows. The specification of the Stochastic Volatility in Mean (SVM) model is given in section 3.2 and in section 3.3 we discuss the financial economic theory on the relationship between returns and volatility. Section 3.4 describes the stock index data and section 3.5 reports on parameter estimation results for the SVM model. In section 3.6 we compare these findings to those obtained with GARCH type models. We also present results for alternative model specifications which we obtain by imposing restrictions on the various parameters in the mean. In the final section we present a summary and some conclusions.

3.2 The Stochastic Volatility in Mean Model

The aim is to simultaneously model the mean and variance of a series of returns on an asset, denoted by y_t. Both the SV and GARCH model are defined by their first and second moment which can be referred to as the mean and variance equation The most general form of the mean equation for both models is then defined as

$$y_t = \mu_t + \sigma_t \varepsilon_t, \qquad \varepsilon_t \sim \text{NID}(0,1)$$
$$t = 1, \ldots, T, \tag{3.1}$$

$$\mu_t = a + \sum_{i=1}^{k} b_i x_{i,t}, \tag{3.2}$$

where the mean μ_t depends on a constant a and regression coefficients $b_1, \ldots, b_k$. The explanatory variables $x_{1,t}, \ldots, x_{k,t}$ may also contain lagged exogenous and dependent variables. The disturbance term ε_t is assumed to be identically and independently distributed with zero mean and unit variance and to this the assumption of normality is added. The positive volatility process is denoted by σ_t and the variance equation for the logarithmic first-order autoregressive SV model is given by

$$
\begin{aligned}
\sigma_t^2 &= \sigma^{*2} \exp(h_t) \\
h_t &= \phi h_{t-1} + \sigma_\eta \eta_t \qquad \eta_t \sim \text{NID}(0,1),
\end{aligned}
\tag{3.3}
$$

where σ^* is a positive scaling factor and h_t is a first-order autoregressive stochastic process with persistence parameter ϕ restricted to a positive value less than one to ensure stationarity. The disturbance terms ε_t and η_{t-s} are assumed to be mutually uncorrelated for all integers s.

For the GARCH$(1,1)$ model the variance equation can be written as

$$
\begin{aligned}
\sigma_t^2 &= \omega + \alpha(y_{t-1} - \mu_{t-1})^2 + \beta\sigma_{t-1}^2 \\
&= \omega + \alpha(\sigma_{t-1}\varepsilon_{t-1})^2 + \beta\sigma_{t-1}^2.
\end{aligned}
\tag{3.4}
$$

with parameter values restricted to $\omega > 0$, $\alpha \geq 0$, $\beta \geq 0$ and $\alpha + \beta < 1$.

The SV model with volatility included in the mean is then given by equations (3.1) and (3.3) where the mean equation (3.2) is rewritten as

$$
\mu_t = a + \sum_{i=1}^{k} b_i x_{i,t} + d\sigma^{*2} \exp(h_t), \tag{3.5}
$$

with d as the regression coefficient measuring the volatility-in-mean effect. In particular, we will use the mean specification

$$
\mu_t = a + b y_{t-1} + d\sigma^{*2} \exp(h_t). \tag{3.6}
$$

This SVM model has six parameters which are to be estimated simultaneously using simulated maximum likelihood estimation methods as discussed in Appendix A which also gives some Monte Carlo evidence of small sample consistency of the estimated parameters. Inclusion of the variance as one of the determinants of the mean facilitates the examination of the relationship between returns and volatility. It enables us to perform studies in the vein of French, Schwert and Stambaugh (1987) but now in the context of SV models. The relative ease with which they were able to conduct their research, *i.e.* without prior manipulation of the original data series, is now also feasible for SV models.

The equivalent in-mean specification for the GARCH model, based on equations (3.1), (3.2) and (3.4), is

$$\mu_t = a + by_{t-1} + d\{\omega + \alpha(\sigma_{t-1}\varepsilon_{t-1})^2 + \beta\sigma_{t-1}^2\}. \tag{3.7}$$

3.3 Some Theory on the Relationship between Returns and Volatility

The relation between expected stock index returns and conditional volatility has received much attention in the financial economic literature. Although a positive relationship between expected returns and expected volatility is consistent with the Capital Asset Pricing Model (CAPM) and intuitively appealing, as rational risk-averse investors require higher expected returns during more volatile periods, empirical research has been unable to establish a convincing positive relationship between the expected risk premium and conditional volatility using GARCH-M models[4]. Instead, there appears to be stronger evidence of a negative relationship between unexpected returns and innovations to the volatility process which French *et al.* (1987) interpret as indirect evidence of a positive correlation between the expected risk premium and *ex ante* volatility. They reason that large pieces of good or bad news induce higher expected volatility for future time periods as volatility is persistent. If expected volatility and expected returns are positively related and future cash flows are unaffected, the current stock index price should fall. Conversely, the arrival of small pieces of news leads to lower expected future volatility and hence to an increase in contemporaneous stock index prices. This theory, known as the volatility feedback theory, therefore hinges on two assumptions. Firstly, volatility persistence and secondly, the existence of a positive relation between the expected components of the return and the volatility process. An alternative explanation for asymmetric volatility where causality runs in the opposite direction is the leverage effect put forward by Black (1976), who asserts that bad (good) news induces a negative (positive) return shock which

[4]See, *e.g.*, for the US stock market French *et al.* (1987) and Campbell and Hentschel (1992), who observe a positive relation, whereas Glosten, Jagannathan and Runkle (1993) present evidence of a negative relation, as does Nelson (1991) with his EGARCH model. Poon and Taylor (1992) who study the UK stock market report a weak positive relationship.

leads to an increase (decrease) in the firm's financial leverage ratio resulting in an upward (downward) effect on the volatility of the stock returns[5]. However, it has been argued by Black (1976), Christie (1982), French *et al.* (1987) and Schwert (1989) that leverage alone cannot account for the magnitude of the negative relationship[6]. If both volatility feedback and leverage effects are present then large pieces of bad news are associated with an increase in volatility, whereas the net impact of large pieces of good news is unclear. The reverse then holds for small shocks to the return process as small positive shocks are associated with a decrease in volatility and the net effect of small negative shocks is unknown.

In the GARCH literature the leverage effect has been empirically established with the EGARCH model of Nelson (1991) and the GJR-GARCH model developed by Glosten *et al.* (1993). In these models the conditional volatility at time $t+1$ is allowed to respond asymmetrically to unanticipated rises and falls in the stock price that occurred at time t. SV models, unlike GARCH models, define volatility as truly contemporaneous and hence its volatility measure includes not only expected but also unexpected volatility. The correlation between unexpected shocks to the return and the volatility process, which can both be observed at time t, could then be measured by estimating $corr(\varepsilon_t, \eta_t)$ as an additional parameter in the SV model[7].

The general mean equation with time-varying variance we consider for estimation is

$$y_t = a + by_{t-1} + d\sigma_t^2 + \sigma_t\varepsilon_t, \tag{3.8}$$

where y_t denotes the excess returns on the stock index at time t and σ_t^2 is the variance which is contemporaneous for the SV model and conditional for the GARCH model. For GARCH models the d parameter therefore measures the relation between expected returns and expected volatility,

[5]Even when the firm has no or little debt it is likely to have operating leverage as its expenses initially remain constant as income falls.

[6]Campbell and Hentschel (1992) find evidence of both volatility feedback and leverage effects, whereas Bekaert and Wu (2000) present results which strongly favour the volatility feedback hypothesis.

[7]Jacquier, Polson and Rossi (2001) estimate $corr(\varepsilon_t, \eta_t)$ and observe a convincing negative relationship between contemporaneous unexpected stock index returns and unexpected volatility. Harvey and Shephard (1996), on the other hand, estimate $corr(\varepsilon_t, \eta_{t+1})$ and Watanabe (1999) develops an SV model which includes the lagged shock to the return process as an explanatory variable in the variance equation allowing for an asymmetric response. Both studies report negative coefficients for the relation between current unexpected returns and *future* volatility.

whereas in the case of SV models the d parameter estimate contains information about the relation between returns on the one hand and expected and unexpected, or *ex post*, volatility on the other. For the GARCH(1,1)-M model the conditional mean is defined as

$$\mathrm{E}_{t-1}(y_t) = a + by_{t-1} + d\{\omega + \alpha(\sigma_{t-1}\varepsilon_{t-1})^2 + \beta\sigma_{t-1}^2\} \tag{3.9}$$

and for the SVM model the equivalent notation is given by

$$\begin{aligned}\mathrm{E}_{t-1}(y_t) &= a + by_{t-1} + d\{\sigma^{*2}\exp(\phi h_{t-1})\exp(\mathrm{E}_{t-1}(\sigma_\eta \eta_t))\} \\ &= a + by_{t-1} + d\{\sigma^{*2}\exp(\phi h_{t-1})\}.\end{aligned} \tag{3.10}$$

It is evident from equations (3.9) and (3.10) that all the explanatory variables in the conditional mean of the GARCH-M(1,1) are known at time $t-1$, whereas the shock to the volatility process at time t, denoted by $\sigma_\eta \eta_t$, prevents observation of the σ_t^2 term in the SVM model at time $t-1$. The mean equation for the SVM model can then be written as

$$\begin{aligned}y_t &= \mathrm{E}_{t-1}(y_t) + d\{\sigma^{*2}\exp(\phi h_{t-1})[\exp(\sigma_\eta \eta_t) - \exp(\mathrm{E}_{t-1}(\sigma_\eta \eta_t))]\} + \\ &\quad \sigma_t \varepsilon_t \\ &= \mathrm{E}_{t-1}(y_t) + d\{\sigma^{*2}\exp(\phi h_{t-1})[\exp(\sigma_\eta \eta_t) - 1]\} + \sigma_t \varepsilon_t,\end{aligned} \tag{3.11}$$

where the second term on the right-hand-side is d multiplied by the *unexpected* volatility at time t. The d coefficient in the SVM model therefore not only measures the relation between the expected components of the return and the volatility process, but also between the unexpected components[8]. This point is clarified when equations (3.10) and (3.11) are combined and written as

$$y_t = a + by_{t-1} + d\sigma_{t|t-1}^2 + d\{\sigma_t^2 - \sigma_{t|t-1}^2\} + \sigma_t \varepsilon_t, \tag{3.12}$$

where $\sigma_{t|t-1}^2$ denotes the conditional variance at time t given the information available at time $t-1$, or the expected volatility, and σ_t^2 the contemporaneous volatility measure at time t. The $\{\sigma_t^2 - \sigma_{t|t-1}^2\}$ term denotes the unexpected shock to the volatility process which should not

[8]Both effects had previously been estimated in the same model by, for example, French *et al.* (1987) and Campbell and Hentschel (1992) who had developed a formal QGARCH-M model of volatility feedback.

be related to the predictable components. The volatility feedback effect is then measured by the d parameter preceding the $\{\sigma_t^2 - \sigma_{t|t-1}^2\}$ term which is expected to be negative as large (small) shocks to the return process raise (lower) contemporaneous volatility through $\sigma_\eta \eta_t$, presumably irrespective of the sign of ε_t, inducing a drop (increase) in the current stock index price in the case of initial large negative (small positive) return shocks which are amplified, whereas initial large positive and small negative shocks are dampened. As deterministic GARCH models do not contain an unexpected volatility component, that is $\sigma_t^2 = \sigma_{t|t-1}^2$, the issue does not arise for this class of volatility models and the d parameter only measures the relation between the expected returns and the expected volatility.

3.4 Data

The data we analyse includes daily stock index returns from three international stock markets: the United Kingdom, the United States and Japan. The UK Financial Times All Share Index and the US Standard & Poor's Composite stock index series cover the period 1 January 1975 to 31 December 1998 whereas the Japanese Topix series starts on 1 January 1988 and ends at 31 December 1998. The stock data was obtained from Datastream. From the same data source we also collected daily UK and Japanese 1-month Treasury bill rates; the US 3-month Treasury bill rate data was extracted from the on-line Federal Reserve Bank of Chicago Statistical Release H.15 database. These interest rate series are used as proxies for the risk free rate of return. The stock index prices are in local currencies and not adjusted for dividends following studies of French *et al.* (1987) and Poon and Taylor (1992), who found that inclusion of dividends affected estimation results only marginally. Returns are calculated on a continuously compounded basis and expressed in percentages, they are therefore calculated as $R_t = 100(\ln P_t - \ln P_{t-1})$ where P_t is the price of the stock market index at time t. From these returns we subtract the daily risk free rate multiplied by 100, denoted by Rf_t, in order to obtain the excess returns which are therefore defined as $y_t = R_t - Rf_t$. In this chapter we model the behaviour of five series: we consider daily excess return series on the UK and US index that cover a period of 24 years ending in 1998, as well as 11-year subsamples of these two series together with excess returns on the Japanese stock market index. These shorter series start in 1988 and therefore exclude the extreme negative observations relating to the 1987 stock market crash. Figure 3.1 contains graphs of the excess return series and the accompanying summary statistics are presented in table 3.1.

Table 3.1. Summary statistics of daily excess returns

Period	1975–1998		1988–1998		
No. of obs. T	6261		2869		
Stock index	FT All	S&P	FT All	S&P	Topix
Mean	0.033	0.028	0.017	0.042	-0.025
Variance	0.943	0.874	0.584	0.747	1.357
Skewness	-0.194	-2.562	-0.022	-0.664	0.343
Excess Kurtosis	11.828	62.758	3.491	7.954	6.107
N	36536	1034312	1458	7774	4516
Excess Returns					
$\hat{\rho}_1$	0.167	0.054	0.115	0.004	0.100
$\hat{\rho}_2$	0.008	-0.024	-0.002	-0.013	-0.062
$\hat{\rho}_3$	0.037	-0.021	-0.005	-0.041	-0.009
$\hat{\rho}_4$	0.046	-0.024	0.041	-0.016	0.027
$\hat{\rho}_5$	0.019	0.032	0.009	0.006	-0.030
$Q(12)$	262.08	41.94	61.32	29.86	64.86
Squared Excess Returns					
$\hat{\rho}_1$	0.478	0.112	0.163	0.176	0.163
$\hat{\rho}_2$	0.281	0.149	0.155	0.087	0.161
$\hat{\rho}_3$	0.238	0.077	0.136	0.049	0.118
$\hat{\rho}_4$	0.290	0.020	0.111	0.087	0.173
$\hat{\rho}_5$	0.202	0.137	0.109	0.097	0.179
$Q(12)$	4543.79	404.14	560.52	286.52	527.91

N is the χ^2 normality test statistic with 2 degrees of freedom; $\hat{\rho}_\ell$ is the sample autocorrelation coefficient at lag ℓ with asymptotic standard error $1/\sqrt{T}$ and $Q(\ell)$ is the Box-Ljung portmanteau statistic based on ℓ squared autocorrelations.

We observe that the effects of the October 1987 crash are especially pronounced for the US stock market where excess returns on the Standard & Poor's Composite index fell by nearly 23% on one single trading day. This one observation contributes to a great extend to the large excess kurtosis value of 62.758 and the high negative skewness coefficient of -2.562. The most volatile series of the five is the Topix series which cannot be attributed to one extreme movement, as can be seen in figure 3.1, but to several prolonged periods of market turbulence initiated in the early nineties by the collapse of the Japanese asset market. The Topix series is further characterised by a negative mean and is positively skewed, which are features not typically found in a stock index (excess)

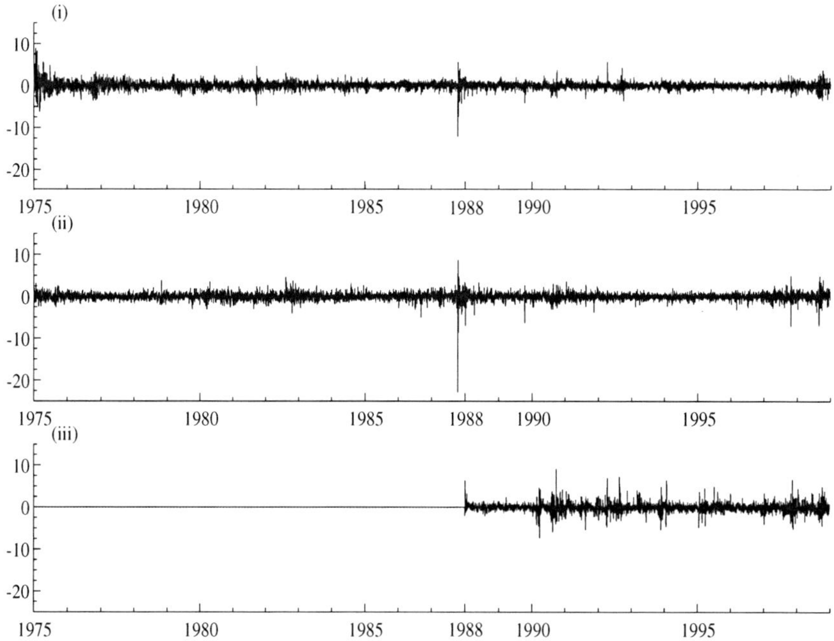

Figure 3.1. Excess returns for the (i) FT All Share Index (UK) and (ii) S&P Composite Stock Index (US) between 02/01/75 and 31/12/98 and for the (iii) Topix Stock Index (Japan) between 04/01/88 and 31/12/98.

return series. We further observe that the UK excess returns and squared excess returns for the period starting in 1975 are highly autocorrelated at lag 1 but that these values are much lower and comparable with those of the Topix stock index for the subsample period 1988–1998. First-order serial correlation coefficients for the Standard & Poor's Composite Index excess returns on the other hand are relatively low for both the full and the subsample period. In the case of excess returns high first-order autocorrelation reflects the effects of non-synchronous or thin trading, whereas highly correlated squared returns can be seen as an indication of volatility clustering. The Q(12) test statistic which is a joint test for the hypothesis that the first twelve autocorrelation coefficients are equal to zero, indicates that this hypothesis has to be rejected at the 1% significance level for all excess return and squared excess return series.

3.5 Estimation Results for the SVM Model and Some Diagnostics

Our main objective in this empirical section is to estimate the contemporaneous relationship between excess returns on stock market indices and their volatility with our SVM model, which we already defined in equations (3.1), (3.3) and (3.6) as

$$\begin{aligned} y_t &= a + by_{t-1} + d\sigma_t^2 + \sigma_t \varepsilon_t, \qquad \varepsilon_t \sim \text{NID}(0,1), \\ \sigma_t^2 &= \sigma^{*2} \exp(h_t), \\ h_t &= \phi h_{t-1} + \sigma_\eta \eta_t, \qquad \eta_t \sim \text{NID}(0,1), \end{aligned}$$

and from equation (3.12) we recall that

$$y_t = a + by_{t-1} + d\sigma_{t|t-1}^2 + d\{\sigma_t^2 - \sigma_{t|t-1}^2\} + \sigma_t \varepsilon_t.$$

Table 3.2 reports the SVM model estimation results for the stock index series over the full sample period 1975-1998 and the subsample period 1988-1998[9]. Volatility persistence estimates for the five series are all highly significant and quite similar with values for ϕ ranging from 0.966 for the post-crash Topix to 0.984 for the full sample Financial Times All Share Index, the series which displayed the highest degree of autocorrelation in the squared returns. This near-unity volatility persistence for high frequency data is consistent with findings from both the SV and the GARCH literature. The two remaining volatility process parameters cover a much wider range. The highest values for the scaling parameter and the parameter which measures the variation in the volatility process are observed for the Topix series with $\sigma^{*2} = 0.832$ and $\sigma_\eta^2 = 0.058$. The SVM model therefore captures the more erratic behaviour of the Topix quite well through a combination of parameters: the high scaling parameter indicates a higher level of volatility whereas the relative small value for ϕ and the large value for σ_η^2 imply that its volatility process is less predictable than that of the other four series.

The three parameters which govern the mean process are reported in the first three rows together with their 95% confidence intervals. We observe that the mean parameter a is always positive and statistically significant for all series with the exception of the Topix series which has a negative sample mean; see table 3.1. We note however that we have

[9]Both the data series and the program used for the estimation of the SVM models which allows for alternative definitions of the mean equation can be found on the Internet at `www.ssfpack.com/holbook`; also see Appendix C.

simultaneously estimated the constant a and the in-mean coefficient d where the latter is associated with a regressor which is strictly positive at all times. The estimates of b are statistically significant for all series and very similar to the first-order autocorrelation coefficients reported in table 3.1.

The d parameter which measures both the *ex ante* relationship between returns and volatility and the volatility feedback effect is negative for all series, although the hypothesis of d equal to zero can never be rejected at the conventional 5% significance level. For some of the series d is however very close to being statistically significant. A similar SVM model was estimated by Fridman and Harris (1998), who studied daily returns on the Standard & Poor's index over the period 1980 to 1987, and Watanabe (1999), who examined the daily Topix series over the eight year period 1990-1997. Both studies reported significant positive values for the contemporaneous relationship[10].

Our findings here are more in line with those of French *et al.* (1987) who regress monthly excess returns of a US stock portfolio against expected and unexpected volatility obtained with ARIMA models based on daily data. For the regression excluding unexpected volatility they observe a weak positive relation between expected returns and volatility. Inclusion of both volatility measures however results not only in a highly significant negative relation between the unexpected components but also turns the sign for the *ex ante* relationship which becomes weakly negative. The negative relation between the unexpected components therefore dominates the weaker, presumably positive, relation between the expected components. We further observe that the largest negative values for d are found for the shorter samples where they are also closest to being statistically significant. A possible explanation is that the positive *ex ante* relation is more pronounced for the period following the stock market crash inducing a stronger volatility feedback effect which forces the d parameter further downward. Such an interpretation is also supported by the findings of Campbell and Hentschell (1992) who incorporate volatility feedback into a GARCH-M model where the volatility feedback parameter is restricted so that a large positive value for the *ex ante* relation implies a large volatility feedback effect[11]. In their study of the US stock market they find that estimates of the unrestricted volatility feedback parameter are very similar to those of the restricted version. French *et al.* (1987) further point

[10]Fridman and Harris (1998) did however not allow for a constant in the mean and the likelihood ratio test for $d = 0$ amounted to a value of 0.14.

[11]Also see Campbell *et al.* (1997, pp. 497-498).

Table 3.2. Estimation results for the SVM model

Period	1975–1998		1988–1998		
No. of obs.	6261		2869		
Stock Index	FT All	S&P	FT All	S&P	Topix
a	0.038	0.045	0.061	0.074	0.019
	0.014 0.063	0.017 0.073	0.016 0.106	0.038 0.111	−0.014 0.052
b	0.146	0.074	0.100	0.024	0.099
	0.123 0.172	0.054 0.101	0.068 0.144	0.009 0.062	0.067 0.143
d	−0.011	−0.023	−0.085	−0.046	−0.031
	−0.049 0.028	−0.065 0.019	−0.176 0.006	−0.103 0.011	−0.066 0.005
σ^{*2}	0.615	0.597	0.458	0.539	0.832
	0.498 0.758	0.508 0.701	0.358 0.587	0.427 0.682	0.642 1.077
ϕ	0.984	0.979	0.976	0.970	0.966
	0.977 0.990	0.969 0.986	0.958 0.986	0.954 0.981	0.947 0.978
σ_η^2	0.018	0.021	0.019	0.035	0.058
	0.014 0.025	0.015 0.029	0.012 0.032	0.025 0.050	0.041 0.082
AIC	15075.6	15005.6	6090.7	6610.7	7988.7
$Q(12)$	24.45	20.06	7.44	21.79	10.30
N	16.265	18.578	2.326	9.969	11.643
σ_∞^2	0.825	0.760	0.560	0.727	1.277

Parameter estimates are reported together with the asymptotic 95% confidence interval which are asymmetric for b, σ^{*2}, ϕ and σ_η^2; AIC is the Akaike Information Criterion which is calculated as -2(ln L) + 2p and $Q(\ell)$ is the Box-Ljung portmanteau statistic for the estimated observation errors which is asymptotically χ^2 distributed with $\ell - p$ degrees of freedom where p is the total number of estimated parameters; N is the χ^2 normality test statistic for ε_t with 2 degrees of freedom; σ_∞^2 denotes the unconditional variance as implied by the volatility process.

out that the volatility feedback effect would be especially strong and dominant when volatilities are highly autocorrelated. As a consequence, rational risk-averse investors require even higher expected returns when unanticipated increases in future volatility are highly persistent. This would be consistent with our findings for the post-crash samples where higher values for ϕ are combined with larger negative values for the in-mean parameter. The relatively large negative estimates for the d parameter we therefore interpret as evidence of the existence of convincing negative feedback effects which appear especially strong when volatility is persistent. This then provides indirect evidence of a positive intertemporal relation between expected excess market returns and its volatility as this is one of the assumptions underlying the volatility feedback hypothesis. Therefore, in the SVM model a negative in-mean parameter indicates that when investors expect higher persistent levels of volatility in the future they require compensation for this in the form of higher expected returns.

With regard to the distributional assumptions we observe that the standardised error term, ε_t, abides the normality assumption only for the Financial Times post-crash period and the $Q(12)$ statistic indicates that little serial correlation remains in the standardised error term. The hypothesis that the first twelve autocorrelation coefficients of ε_t are equal to zero can also not be rejected for the Topix series as the critical value at the 5% significance level is 12.6. Even though the remaining values for the $Q(12)$ and normality statistics exceed their critical values, they are much smaller than those observed in table 3.1, especially the values for the normality statistic are substantially reduced.

In addition to the SVM model we also estimated a number of restricted versions of our model which were obtained by restricting one or more of the mean parameters to be equal zero. For the estimation results of $d = 0$ and $a = b = d = 0$ we refer to tables 3.3 and 3.4, respectively. The likelihood ratio tests for the hypothesis $H_0 : d = 0$ never exceeded the critical χ_1^2 5% significance value of 3.84, confirming the insignificance of the d parameter. The estimates of the remaining parameters for this restricted model only changed marginally with the exception of the a parameter which decreased in value. More importantly, however, the likelihood ratio tests for the null hypothesis $H_0 : a = b = d = 0$ were all significant at the 1% significance level. This was confirmed by the values of the AIC statistic, which is a goodness-of-fit statistic that allows for comparison between models with different numbers of parameters.

Table 3.3. Estimation results for the SVM model with $d = 0$

Period	1975–1998		1988–1998		
No. of obs.	6261		2869		
Stock Index	FT All	S&P	FT All	S&P	Topix
a	0.033	0.033	0.027	0.055	−0.0001
	0.016 0.050	0.017 0.049	0.004 0.050	0.032 0.078	−0.026 0.025
b	0.146	0.075	0.101	0.025	0.100
	0.122 0.174	0.054 0.102	0.069 0.146	0.010 0.060	0.088 0.113
σ^{*2}	0.615	0.597	0.459	0.540	0.831
	0.503 0.752	0.504 0.706	0.358 0.588	0.429 0.680	0.734 0.941
ϕ	0.984	0.979	0.976	0.970	0.965
	0.977 0.989	0.970 0.985	0.959 0.986	0.954 0.981	0.950 0.976
σ_η^2	0.018	0.020	0.019	0.035	0.059
	0.014 0.024	0.015 0.028	0.012 0.031	0.022 0.055	0.044 0.079
$LR(d=0)$	0.25	1.13	2.87	1.89	1.95
AIC	15073.8	15002.7	6091.6	6610.6	7988.6
$Q(12)$	24.43	19.88	7.60	21.26	10.42
N	16.855	20.261	3.989	11.936	8.671
σ_∞^2	0.825	0.761	0.562	0.729	1.278

Parameter estimates are reported together with the asymptotic 95% confidence interval which are asymmetric for b, σ^{*2}, ϕ and σ_η^2; $LR(d = 0)$ is the likelihood ratio statistic for the hypothesis $d = 0$; AIC is the Akaike Information Criterion which is calculated as -2(ln L) + 2p and $Q(\ell)$ is the Box-Ljung portmanteau statistic for the estimated observation errors which is asymptotically χ^2 distributed with $\ell - p$ degrees of freedom where p is the total number of estimated parameters; N is the χ^2 normality test statistic with 2 degrees of freedom; σ_∞^2 denotes the unconditional variance as implied by the volatility process.

Table 3.4. Estimation results for the SVM model with $a = b = d = 0$

Period	1975–1998		1988–1998		
No. of obs.	6261		2869		
Stock Index	FT All	S&P	FT All	S&P	Topix
σ^{*2}	0.631	0.602	0.465	0.546	0.839
	0.502 0.793	0.509 0.713	0.365 0.592	0.424 0.703	0.654 1.076
ϕ	0.985	0.979	0.976	0.971	0.965
	0.980 0.989	0.971 0.985	0.962 0.985	0.948 0.984	0.946 0.977
σ_η^2	0.017	0.020	0.019	0.033	0.059
	0.014 0.022	0.014 0.027	0.012 0.031	0.019 0.058	0.041 0.085
$LR(a = b = d = 0)$	144.5	50.1	36.8	24.0	28.6
AIC	15214.1	15049.7	6121.6	6628.7	8011.3
$Q(12)$	151.76	55.40	35.49	23.02	39.77
N	18.422	31.913	4.503	18.788	4.914
σ_∞^2	0.849	0.765	0.570	0.732	1.288

Parameter estimates are reported together with the asymptotic 95% confidence interval which are asymmetric for σ^{*2}, ϕ and σ_η^2; $LR(a = b = d = 0)$ is the likelihood ratio statistic for the hypothesis $a = b = d = 0$; AIC is the Akaike Information Criterion which is calculated as -2(ln L) + 2p and $Q(\ell)$ is the Box-Ljung portmanteau statistic for the estimated observation errors which is asymptotically χ^2 distributed with $\ell - p$ degrees of freedom where p is the total number of estimated parameters; N is the χ^2 normality test statistic with 2 degrees of freedom; σ_∞^2 denotes the unconditional variance.

Therefore we conclude that the SVM model is a useful empirical model for the modelling of the relation between current returns and returns of the previous period together with the feedback effect of current volatility. An additional advantage of modelling the mean is that ε_t behaved better, especially in terms of the assumption of zero autocorrelation.

3.6 Some Comparisons with GARCH-M Estimation Results

The GARCH model we estimate is the GARCH-M(1,1) model defined in equations (3.1), (3.4) and (3.7) as

$$\begin{aligned} y_t &= a + by_{t-1} + d\sigma_t^2 + \sigma_t\varepsilon_t, \qquad \varepsilon_t \sim \text{NID}(0,1), \\ \sigma_t^2 &= \omega + \alpha(\sigma_{t-1}\varepsilon_{t-1})^2 + \beta\sigma_{t-1}^2. \end{aligned}$$

As research in the empirical GARCH literature has shown that the assumption of normally distributed error terms is often violated we estimate the GARCH-M(1,1) model in this section with ε_t following a Student-t instead of a Gaussian distribution[12]. Given our results in the previous section the need to impose an alternative error distribution for the SV class of volatility models is less evident and attributable to the fact that SV models are by definition better suited to incorporate extreme values[13].

The estimation results for the GARCH-M model are given in table 3.5 where we observe estimates for the b parameter that are very similar to those obtained with the SVM model and that near-zero estimates for a are combined with positive values for the d parameter. The null hypothesis of a zero *ex ante* relationship between excess returns and volatility can however never be rejected at the 5% significance level, although in some cases only by a small margin. With regard to our estimation results we further observe that for the post-crash samples the relationship is strongest. In fact, the magnitude of the estimates for d are quite similar to those obtained with the SVM model in absolute terms, so a large negative contemporaneous relationship in the SVM model is accompanied by a large positive *ex ante* relationship in the GARCH-M model. This could be interpreted as confirmation of our hypothesis that a stronger *ex ante* relationship between the return and volatility process

[12] We also estimated the GARCH-M model with a normally distributed error term and observed extremely high values for the normality statistic.

[13] For SV models with alternative error distributions see: Fridman and Harris (1998), Sandmann and Koopman (1998) and Liesenfeld and Jung (2000).

Table 3.5. Estimation results for the GARCH-M model with t-distribution

Period	1975–1998		1988–1998		
No. of obs.	6261		2869		
Stock Index	FT All	S&P	FT All	S&P	Topix
a	0.023 −0.006 0.053	0.007 −0.026 0.041	0.000 −0.052 0.051	0.024 −0.021 0.070	−0.032 −0.079 0.015
b	0.145 0.120 0.170	0.061 0.037 0.086	0.094 0.057 0.132	0.021 −0.032 0.037	0.085 0.047 0.122
d	0.017 −0.024 0.058	0.038 −0.010 0.086	0.062 −0.039 0.162	0.049 −0.021 0.120	0.028 −0.017 0.072
ω	0.014 0.009 0.019	0.008 0.005 0.011	0.009 0.003 0.015	0.004 0.001 0.007	0.028 0.015 0.041
α	0.084 0.070 0.099	0.043 0.034 0.052	0.055 0.037 0.074	0.034 0.022 0.046	0.108 0.080 0.135
β	0.897 0.880 0.914	0.947 0.936 0.958	0.928 0.903 0.953	0.962 0.950 0.975	0.876 0.848 0.904
v	11.914 10.342 14.049	6.046 5.360 6.933	8.704 6.987 11.538	4.572 3.919 5.487	4.955 4.152 6.142
$\alpha + \beta$	0.982	0.990	0.984	0.996	0.983
$Q(12)$	38.53	16.16	11.81	19.53	14.42
σ^2_∞	0.792	0.775	0.561	0.985	1.689

Parameter estimates are reported together with the asymptotic 95% confidence interval which are all symmetric with the exception of those of the v parameter; $Q(\ell)$ is the Box-Ljung portmanteau statistic for the estimated observation errors which is asymptotically χ^2 distributed with $\ell - p$ degrees of freedom where p is the total number of estimated parameters; σ^2_∞ denotes the unconditional variance as implied by the volatility process.

Table 3.6. Estimation results for the GARCH-M model with $d = 0$ and t-distribution

Period	1975–1998		1988–1998		
No. of obs.	6261		2869		
Stock Index	FT All	S&P	FT All	S&P	Topix
a	0.033 0.015 0.051	0.030 0.012 0.048	0.027 0.004 0.051	0.051 0.027 0.075	−0.009 −0.039 0.021
b	0.145 0.120 0.170	0.062 0.037 0.086	0.095 0.057 0.133	0.003 −0.032 0.037	0.087 0.050 0.124
ω	0.014 0.009 0.019	0.008 0.004 0.011	0.009 0.003 0.015	0.004 0.001 0.007	0.028 0.015 0.041
α	0.084 0.070 0.099	0.043 0.034 0.052	0.055 0.037 0.074	0.033 0.022 0.045	0.107 0.080 0.134
β	0.897 0.880 0.914	0.947 0.937 0.958	0.929 0.904 0.954	0.963 0.951 0.976	0.876 0.848 0.904
υ	11.904 10.332 14.040	6.046 5.360 6.934	8.699 6.984 11.529	4.569 3.917 5.483	4.930 4.135 6.104
$\alpha + \beta$	0.982	0.990	0.984	0.996	0.983
$LR(d = 0)$	0.68	3.72	1.50	2.06	1.68
$Q(12)$	38.35	16.19	11.32	19.86	13.23
σ^2_∞	0.792	0.777	0.562	0.994	1.690

Parameter estimates are reported together with the asymptotic 95% confidence interval which are all symmetric with the exception of those for the υ parameter; $LR(d = 0)$ is the likelihood ratio statistic for the hypothesis $d = 0$; $Q(\ell)$ is the Box-Ljung portmanteau statistic for the estimated observation errors which is asymptotically χ^2 distributed with $\ell - p$ degrees of freedom where p is the total number of estimated parameters; σ^2_∞ denotes the unconditional variance as implied by the volatility process.

Table 3.7. Estimation results for the GARCH-M model with $a = b = d = 0$ and t-distribution

Period	1975–1998		1988–1998		
No. of obs.	6261		2869		
Stock Index	FT All	S&P	FT All	S&P	Topix
ω	0.013	0.007	0.009	0.004	0.028
	0.009 0.018	0.004 0.010	0.003 0.015	0.001 0.006	0.015 0.041
α	0.080	0.041	0.055	0.032	0.104
	0.066 0.094	0.032 0.050	0.037 0.073	0.021 0.043	0.078 0.131
β	0.903	0.950	0.929	0.964	0.879
	0.887 0.920	0.940 0.961	0.904 0.954	0.952 0.976	0.852 0.907
υ	11.290	5.900	8.448	4.547	4.843
	9.872 13.183	5.228 6.768	6.803 11.142	3.886 5.480	4.061 5.999
$\alpha + \beta$	0.983	0.991	0.984	0.997	0.983
$LR(a = b = d = 0)$	154.90	47.90	38.52	30.38	28.20
$Q(12)$	208.14	59.29	49.38	21.37	45.83
σ^2_∞	0.812	0.787	0.569	1.040	1.688

Parameter estimates are reported together with the asymptotic 95% confidence interval which are all symmetric with exception of those for the υ parameter; $LR(a = b = d = 0)$ is the likelihood ratio statistic for the hypothesis $a = b = d = 0$; $Q(\ell)$ is the Box-Ljung portmanteau statistic for the estimated observation errors which is asymptotically χ^2 distributed with $\ell - p$ degrees of freedom where p is the total number of estimated parameters; σ^2_∞ denotes the unconditional variance as implied by the volatility process.

induces a more convincing volatility feedback effect which dominates the relation between expected returns and expected volatility.

The volatility persistence parameters are comparable to those found for the SVM model with near-unity values for the sum of α and β, although we observe that the persistence values for the Topix and the Standard & Poor's sample starting in 1988 are considerably higher when modelled with the GARCH-M model. They are in fact so high that they exceed those of the 1975 – 1998 Financial Times All Share Index series which exhibits very high autocorrelated squared returns as shown in table 3.1. We find that the GARCH-M model also captures the volatility in the variance equation of the Topix series as the α parameter, which measures the extent to which volatility at time t is influenced by a shock to the return process at time $t-1$, is highest for this series.

Imposing restrictions of the form $d = 0$ and $a = b = d = 0$ were also carried out for the GARCH-M model and results can be found in tables 3.6 and 3.7, respectively. Again the estimates of the remaining parameters only changed marginally, with the exception of the a parameter which increased when we set d equal to zero. The likelihood ratio statistic indicated that the in-mean effect had little explanatory value as the hypothesis of $d = 0$ could not be rejected at the 5% significance level for any of the five samples. The hypothesis $a = b = d = 0$ on the other hand had to be rejected for all series, which was consistent with our findings for the SVM model.

3.7 Summary and Conclusions

In this chapter we have presented a Stochastic Volatility model where the mean is modelled simultaneously with the variance equation. When one of the variables in the mean is the volatility process itself, we obtain the Stochastic Volatility in Mean (SVM) model with which we are able to investigate the contemporaneous relationship between excess returns on a stock market index and its time-varying volatility.

For the empirical application we examined stock indices from the United Kingdom, the United States and Japan over two time periods. The results were then compared with the estimation results obtained for their GARCH counterparts. The conclusions of our empirical study can be summarised as follows. Firstly, with our SVM model we find evidence of a weak negative relationship for all stock index series, whereas estimation with the GARCH-M model produces statistically insignificant positive estimates for the in-mean parameter. The difference in the sign for d is directly attributable to the fact that the SVM model, unlike the GARCH-M model, not only measures the relation between expected returns and expected volatility but also the volatility feedback effect where the latter appears to dominate the former. The largest negative values are then observed for those series which have a larger positive d estimate in the GARCH-M models and we assert that a strong *ex ante* positive relation induces a more convincing volatility feedback effect which then provides indirect evidence of the positive *ex ante* relationship between the return and the volatility process as this is one of the main assumptions underlying the volatility feedback hypothesis. Secondly, we find that simultaneous modelling of the mean and the variance equation lead to a better fit of the volatility series. The first-order autoregressive term b in the mean equation appears robust across model specifications and classes of volatility models. Although it is possible to model the original series prior to estimation with a volatility model, simultaneous estimation is more efficient. Finally, we observe that the volatility persistence parameter ϕ in the SV models, which is an indication of volatility clustering, is comparable with the persistence measure $(\alpha + \beta)$ of GARCH models. An advantage of SV models over GARCH models is that the distributional assumptions of the error term in the mean ε_t are much less violated. This makes the case for alternative error distributions and hence the estimation of an additional parameter less strong for the SV class of volatility models. On the basis of the above we therefore feel that SV models can be regarded as a competitive alternative to GARCH models, not only in theoretical terms but also in empirical research.

Chapter 4

FORECASTING WITH VOLATILITY MODELS

Forecasting of financial market volatility has been a major interest to both academic researchers and practitioners for many years. This level of interest is justified as the relevance of volatility forecasts is felt in many areas of financial decision making such as portfolio allocation, risk management, hedging strategies and derivatives pricing. Extensive research in this field has resulted in the proposal and development of a variety of models and procedures which are either motivated by the notion that (1) option prices contain valuable information about the future volatility of the underlying asset or that (2) historical financial market volatility calculated from fitting an appropriate volatility model can be successfully cast into the future. In Chapter 5 we further discuss the concept of implied volatility whereas in this chapter we focus on the forecasting ability of volatility models based on historical price information. The remaining sections of this chapter are then organised as follows. In the next section we show how the forecasts of the various volatility models evolve over time with emphasis on the Stochastic Volatility (SV) and Generalised Autoregressive Conditional Heteroskedasticity (GARCH) classes of volatility models. In section 4.2 we then present an empirical out-of-sample forecasting study of six international stock indices based on four different volatility models and for different forecasting horizons.

4.1 Volatility Models and Their Forecasts

Empirical evidence accumulated over the past few decades has unambiguously shown that volatility is a time-varying process. Whereas it is generally acknowledged that SV and GARCH models most adequately describe the empirical in sample volatility behaviour of financial time series, there is still no consensus as to which model is the most accurate in terms of out-of-sample forecasting ability[1]. Prior to the introduction of (G)ARCH models by Engle (1982) and Bollerslev (1986), which marked the beginning of a research era where time-variation in volatility was modelled explicitly, less "sophisticated" historical models were used for the purpose of volatility prediction[2]. The simplest of these is the random walk (RW) model with volatility calculated as the average of squared returns over the sample period[3].

In empirical applications with daily stock index return data the persistence parameter estimates of both SV and GARCH models are invariably found to be close to unity which suggests that the volatility structure of financial time series at higher sampling frequencies evolves slowly over time. A priori we therefore expect volatility predictions based on these time-varying volatility models to be fairly accurate. The near-unity estimates for ϕ and $(\alpha+\beta)$ imply that for shorter forecasting horizons the value of the long-term volatility forecast is of second-order importance and that forecasts are mainly determined by the size of the short-term volatility forecast, denoted by $\hat{\sigma}^{*2}\exp(\hat{h}_{T+1|T}+0.5p_{T+1|T})$ for the SV model and by $\hat{\sigma}^2_{T+1|T}$ for GARCH models. In the SV model $\hat{\sigma}^{*2}$ is the maximum likelihood estimate of σ^{*2} and $\hat{h}_{T+1|T}$ denotes the estimator of h_{T+1} using all observations T with variance $p_{T+1|T}$[4]. For the more than one-period-ahead forecasts these two terms are given by $\hat{h}_{T+j|T} = \hat{\phi}^{j-1}\hat{h}_{T+1|T}$ and $p_{T+j|T} = \hat{\phi}^{2(j-1)}p_{T+1|T} + \sum_{i=0}^{N-2}\hat{\phi}^{2i}\hat{\sigma}^2_\eta$ with $j \geq 2$ and where $\hat{\phi}$ and $\hat{\sigma}^2_\eta$ are the maximum likelihood estimates of ϕ and σ^2_η, respectively. Short-term volatility in the GARCH(1,1) can be calculated directly as all necessary information is known at time T and for $j \geq 2$ we obtain $\hat{\sigma}^2_{T+j|T} = \hat{\omega}/(1-\hat{\alpha}-\hat{\beta}) + (\hat{\alpha}+\hat{\beta})^{j-1}(\hat{\sigma}^2_{T+1|T} - \hat{\omega}/(1-\hat{\alpha}-\hat{\beta}))$ where $\hat{\omega}$, $\hat{\alpha}$ and $\hat{\beta}$ are the maximum likelihood estimates of ω, α and β, respectively. It is evident that for very large values of j the long-term volatility forecasts, which are identical to the unconditional variances, can be written as $\hat{\sigma}^{*2}\exp(0.5\hat{\sigma}^2_\eta/(1-\hat{\phi}^2))$ for the SV model and as $\hat{\omega}/(1-\hat{\alpha}-\hat{\beta})$ for the GARCH(1,1) model where the rate of convergence depends on the respective persistence estimates ϕ and $(\alpha+\beta)$.

---

[1] Model specifications of the SV and GARCH models are given in Chapter 2.

[2] See, for example, Brailsford and Faff (1996) and Walsh and Tsou (1998).

[3] We do not deduct the sample mean following Figlewski (1997), who for smaller samples finds the sample mean to be a very noisy estimate of the true mean.

[4] These values are computed by simulation methods as discussed in Appendix A. As the Kalman filter yields $\hat{h}_{T+1|T}$ and $p_{T+1|T}$ this enables us to work with these values instead of $\hat{h}_{T|T}$ and $p_{T|T}$; see Harvey (1989, Chapter 3).

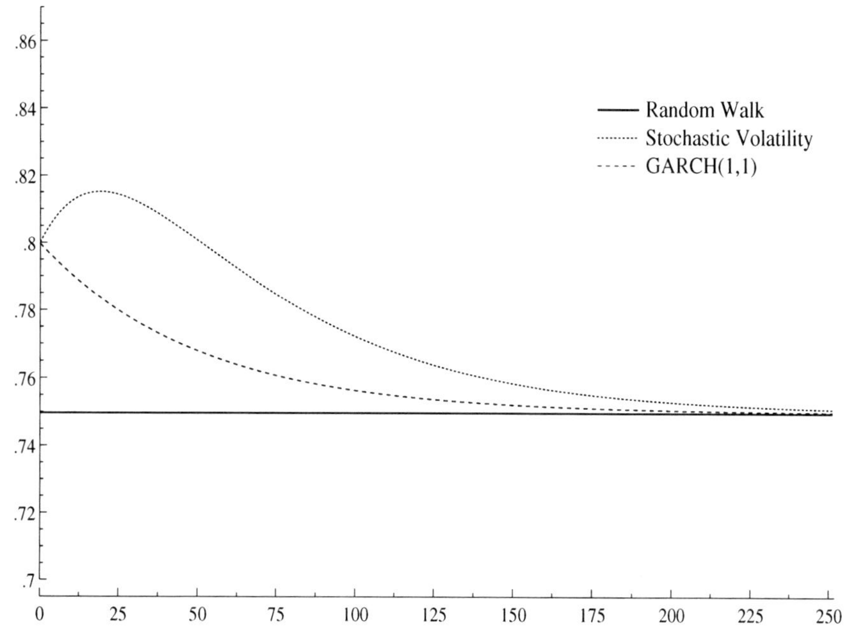

Figure 4.1. Volatility forecasts produced by the RW, SV and GARCH(1,1) model. RW, SV and GARCH(1,1) model volatility forecasts with persistence parameters $\hat{\phi}$ and $(\hat{\alpha}+\hat{\beta})$ set to 0.98 and for $j = 1, \ldots, 252$; long-term volatility forecast values are fixed at 0.75 for all three models.

In figure 4.1 we plot three theoretical volatility forecasting series which are based on the RW, SV and GARCH(1,1) models for forecasting horizons $j \geq 1$. Selected parameter values reflect estimates typically found in empirical studies for daily stock index return series with $\hat{\phi} = 0.98$ for the SV and $\hat{\alpha} + \hat{\beta} = 0.98$ for the GARCH model. In addition we select an unconditional variance or long-term volatility forecast value of 0.75 for all three models. In the case of the GARCH model this implies $\hat{\omega} = 0.015$ and for the SV model we set $\hat{\sigma}_\eta^2 = 0.016$ and $\hat{\sigma}^{*2} = 0.613$. We further assume that recent volatility has been higher than 0.75 and we fix the value for the first volatility forecast at 0.80, *i.e.* for the GARCH model $\hat{\sigma}^2_{T+1|T} = 0.80$ and for the SV model we impose the combination $\hat{h}_{T+1|T} = 0.247$ and $p_{T+1|T} = 0.04$. The RW model has a constant volatility measure and hence its forecasts do not change over time.

With a calendar year consisting of approximately 252 trading days we plotted the series for $j = 1, \ldots, 252$ and observe that the SV and GARCH forecasts require almost a full year to converge to the unconditional variance value of 0.75. Given the relatively small difference between short and long-term volatility of 0.05 this is a remarkably long period. The graph therefore illustrates the point we made above about persistence parameters close to unity and shorter forecasting horizons, *i.e.* for small values of j short-term volatility is of greater importance than long-term volatility.

The most notable difference between the forecasts of the time-varying volatility models is that the GARCH forecasting line smoothly decays towards its long-term volatility forecast value whereas the SV line shows a hump before it declines and starts to converge. This initial increase is due to relatively large values for $0.5\sum_{i=0}^{j-1}\hat{\phi}^{2i}\hat{\sigma}_\eta^2$ which is the second part in the error-correction term of the SV forecasting equation. For small values of j the increase in value of this part of the forecasting equation exceeds the decrease in predicted volatility due to $\hat{\phi}^j\hat{h}_{T+1|T}+0.5\hat{\phi}^{2j}p_{T+1|T}$. The size of the observed hump depends on all estimated values in the SV model with the exception of the scaling parameter $\hat{\sigma}^{*2}$ but it disappears altogether when, ceteris paribus, *e.g.* $\hat{\phi}=0.97$ or $\hat{\sigma}_\eta^2=0.011$. Volatility forecasts produced by the SV model are therefore quite sensitive to the estimated input parameters which can, or cannot, result in the curvature shown in figure 4.1. The above also illustrates the difference between the SV and the GARCH model in terms of complexity: whereas the GARCH volatility forecasts are defined as an exponentially decaying mean reverting function, SV model forecasts are constructed in a much more elaborate way. Firstly, the volatility process of the SV model is expressed in logarithmic form which necessitates the use of an error-correction term in the forecasting equation. Secondly, the definition for h_t includes an error term η_t and its variance $\hat{\sigma}_\eta^2$ is an important input parameter in the SV model forecasting formula especially for larger values of j.

4.2 An Empirical Study of Six International Stock Indices

The comparative accuracy with which volatility models are able to predict the variation in returns of stock market indices has been the subject of many empirical studies. The introduction of the GARCH class of volatility models further stimulated this line of research and Akgiray (1989) was the first to compare its volatility forecasting performance to that of the thus far employed models. His findings signalled the potential dominance of GARCH models in this area as US stock index forecasts obtained with the GARCH(1,1) model always outperformed those of the three other models which were the RW, the exponentially weighted moving average (EWMA) and the ARCH model. In the following decade research was extended to other stock indices and more elaborate GARCH models were considered[5]. The conclusions that emerged from these subsequent studies proved to be less consistent than suggested by Akgiray's original results. Tse and Tung (1992), for example, preferred the EWMA model to the GARCH model specifications for the Singapore stock index. Brailsford and Faff (1996) were hesitant to suggest a single superior model for the Australian stock market yet cautiously concluded that evidence seemed to favour the asymmetric Glosten, Jagannathan and Runkle (GJR) GARCH model. German, Dutch, Spanish, Italian and Swedish stock market volatility forecasts were compared by Franses and van Dijk (1996) who in addition to the RW, GARCH and GJR-GARCH model also considered the Quadratic GARCH specification due to Engle and Ng (1993). They preferred the RW model to the asymmetric QGARCH model when the estimation sample included extreme events such as the 1987 stock market crash but reached the opposite conclusion when the sample no longer contained these observations[6]. The GJR-GARCH model on the other hand always performed very poorly. It has been suggested that these contradictory results might stem from the fact that GARCH models suffer from convergence problems as they require a larger number of observations than the simpler methods and this would make GARCH models less suitable for volatility forecasting with smaller sample sizes and lower sampling frequencies; see, for example, Figlewski (1997) and Walsh and Tsou (1998). Poon and Granger (2001a) however pointed out that as volatility in the GARCH models is separated into volatility due to recent shocks and volatility due to the persistence structure of the model, it is more adaptive than that of the less advanced models such as the EWMA model. As SV models share this feature with GARCH models it is of interest to investigate their forecasting performance. Such a study was undertaken by Heynen (1995) who compared the SV model to the RW, GARCH and EGARCH models for the US, Japanese, UK, French, Dutch, Australian and Hong Kong stock market based on estimation samples consisting of eight years of daily return data. He observed that the SV model consistently outperformed the other models and his findings were confirmed by Yu (2000) for the New Zealand stock market. These results were however not repeated for foreign exchange markets as shown by Heynen (1995) and Dunis and Laws (2000). With regard to the stock market studies it

[5]An overview of this research is provided in, for example, Poon and Granger (2001a) who list a total 37 papers on the subject of volatility forecasting with historical price information.
[6]Also see: McMillan, Speight and ap Gwilym (2000).

is further noted that Heynen (1995) based his daily volatility forecasts on monthly updated estimation samples which always included 1987 stock market crash data and that Yu (2000) only constructed *monthly* volatility forecasts from daily time-varying volatility models. We therefore conduct an empirical study for six international daily stock indices in the remainder of this section in order to investigate further the relative forecasting performance of four different volatility models. The models we examine are the RW, EWMA, SV and GARCH models.

4.2.1 Data and Methodology

Data

The stock index series we selected are the UK Financial Times All Share, the US Standard & Poor's Composite, the Japanese Topix, the German DAX, the French CAC 40 and the Dutch AEX. The data were obtained from Datastream and span the twelve-year period of 1 January 1988 to 31 December 1999; each of these series consists of 3131 daily observations. Returns are calculated as the first differential of the natural logarithms of stock market prices on consecutive days and expressed in percentages terms so $y_t = 100(\ln P_t - \ln P_{t-1})$, where P_t denotes the closing price of the stock index at time t.

The daily return series are graphed for all six stock indices over the full sample period in figure 4.2 together with the accompanying histograms and correlograms of the first 30 sample autocorrelation coefficients for the squared returns. Table 4.1 gives the summary statistics. From the graphs we can deduce that the Financial Times All Share Index is the least volatile of the six stock indices and that it has relatively few outliers. Also, the sample skewness coefficient is not significantly different from zero at very high confidence levels with a t-statistic value of 0.66, although excess kurtosis is clearly present[7]. The DAX has the highest degree of variation with variance 1.521 which is mainly due to a few negative outliers. These outliers also largely explain the excess kurtosis value of 9.977 and the negative skewness coefficient of -0.784. Although the much documented diversion from normality cannot be rejected for any of the series, it follows from the histograms in combination with the sample statistics for the third and fourth moment that the assumption of alternative error distributions for the SV and GARCH models is not directly apparent especially as these models are by definition capable of

[7]For normally distributed data the standard error of the skewness and kurtosis estimates are given by $\sqrt{6/T}$ and $\sqrt{24/T}$, respectively.

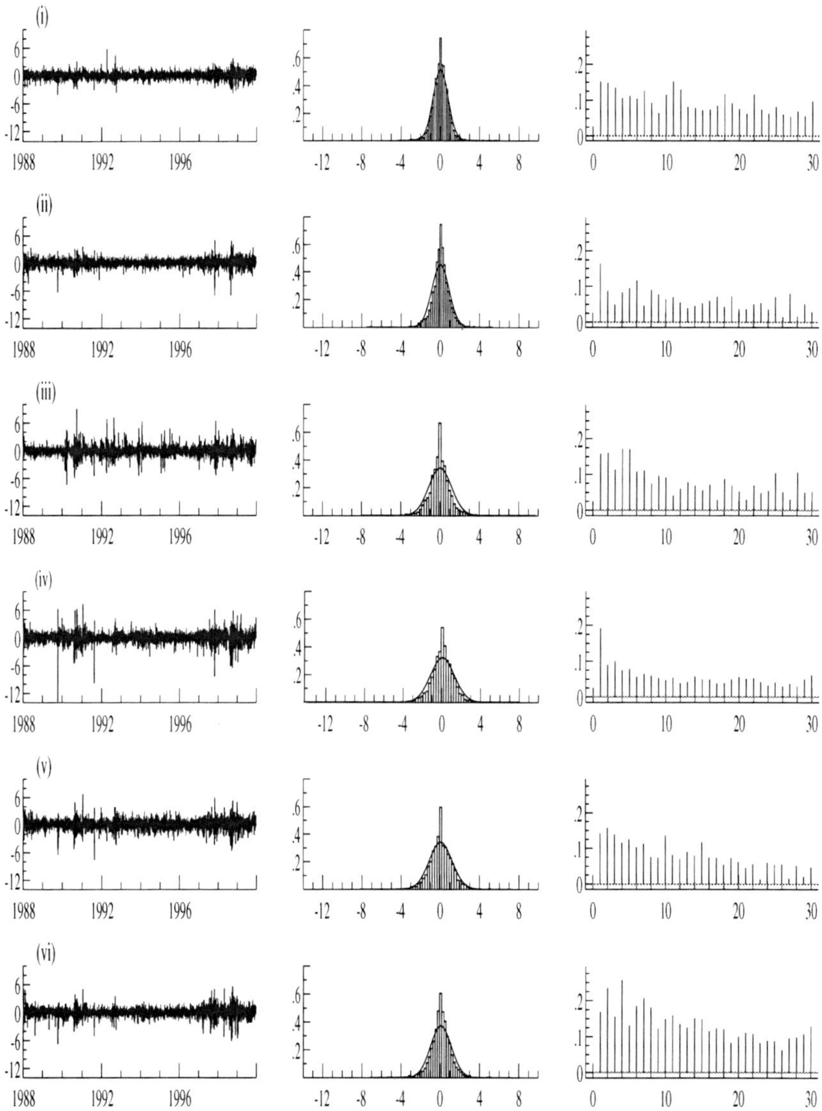

Figure 4.2. Daily returns on the (i) FT All Share, (ii) S&P Composite, (iii) Topix, (iv) DAX, (v) CAC40 and (vi) AEX Stock Index over the period 04/01/88 to 31/12/99 For each stock index return series the time series is plotted on the left, the histogram in the centre and the sample autocorrelation coefficients $\hat{\rho}_\ell$ for $\ell = 1, \ldots, 30$ of the squared returns, on the right.

Table 4.1. Summary statistics of daily stock index returns

Period	1988 - 1999					
No. of obs T	3130					
Stock index	FT All	S&P	Topix	DAX	CAC40	AEX
Mean	0.042	0.057	-0.000	0.062	0.057	0.069
Variance	0.612	0.788	1.354	1.521	1.371	1.140
Skewness	-0.029	-0.546	0.303	-0.784	-0.168	-0.310
Exc.Kurtosis	3.032	6.627	5.674	9.977	2.878	4.855
Maximum	5.698	4.989	9.116	7.288	6.808	6.636
Minimum	-4.147	-7.113	-7.365	-13.71	-7.573	-6.792
$\hat{\rho}_1$	0.153	0.166	0.157	0.192	0.142	0.168
$\hat{\rho}_2$	0.149	0.088	0.160	0.090	0.158	0.235
$\hat{\rho}_3$	0.135	0.049	0.113	0.099	0.139	0.155
$\hat{\rho}_4$	0.106	0.084	0.171	0.075	0.116	0.257
$\hat{\rho}_5$	0.112	0.095	0.170	0.078	0.126	0.131
$Q(12)$	569.1	304.2	543.2	272.5	496.0	1197.3
$Q(30)$	940.6	455.2	783.2	390.1	721.7	1881.9

$\hat{\rho}_\ell$ is the sample autocorrelation coefficient at lag ℓ for the squared returns with asymptotic standard error $1/\sqrt{T}$ and $Q(\ell)$ is the Box-Ljung portmanteau statistic based on ℓ squared autocorrelations with critical values at the 1 percent level for $\ell = 12$ and 30 of 26.22 and 50.89, respectively.

absorbing a certain degree of excess kurtosis[8]. It appears from the graphs that in recent years some stock markets have become more volatile. The correlograms of the squared returns then indicate that volatility clustering is present in all six return series but that it is most prominent for the AEX index followed by the Financial Times All Share Index. This is confirmed by the statistics in table 4.1 which also show that the $Q(\ell)$ values are lowest for the DAX and the Standard & Poor's series. As a high degree of volatility persistence suggests a more predictable volatility process we might expect more accurate volatility forecasts for the AEX and the Financial Times All Share than for the DAX and Standard & Poor's Composite Index.

[8]Whereas model estimation with fat-tailed distributions often results in an improved in-sample fit, research by Claessen and Mittnik (1996) and Peters (2001) has shown that the use of non-normal distributions for out-of-sampling forecasting purposes is far less evident.

Methodology

The models we consider for our forecasting study are the RW, EWMA, SV and GARCH model with volatility forecasts expressed in terms of variances[9]. As the RW model is the only method which does not recognise the fact that volatility is a time-varying process it is mainly included for benchmark purposes. The RW forecasts are given by

$$\hat{\sigma}^2_{T+1,T+N|T} = \frac{N}{n}\sum_{t=0}^{n-1} y^2_{T-t}, \tag{4.1}$$

where N denotes the length of the forecasting horizon and n is the length of the historical sample[10]. The EWMA model is a combination of an exponential smoothing and moving average model and forecasts of this model are obtained by

$$\hat{\sigma}^2_{T+1,T+N|T} = N\left[(1-\hat{\phi})\hat{\sigma}^2_{T|T} + \hat{\phi}\frac{1}{L}\sum_{t=0}^{L-1} y^2_{T-t}\right], \tag{4.2}$$

where $0 < \hat{\phi} \leq 1$ is the smoothing factor. The EWMA model is similar to an exponential smoothing model where volatility of the last period, *i.e.* y^2_T, is replaced with a moving average term based on the previous L squared returns; we choose L so that the second part in equation (4.2) represents a one-month moving average forecast. Section 4.1 provided the forecasting formulas of the SV and the GARCH(1,1) model and for forecasting horizon N these are written as

$$\hat{\sigma}^2_{T+1,T+N|T} = \hat{\sigma}^{*2}\sum_{j=1}^{N} \exp(h_{T+j|T} + 0.5p_{T+j|T}), \tag{4.3}$$

and

$$\hat{\sigma}^2_{T+1,T+N|T} = \sum_{j=1}^{N}\left[\frac{\hat{\omega}}{1-\hat{\alpha}-\hat{\beta}} + (\hat{\alpha}+\hat{\beta})^{j-1}\left(\hat{\sigma}^2_{T+1|T} - \frac{\hat{\omega}}{1-\hat{\alpha}-\hat{\beta}}\right)\right], \tag{4.4}$$

[9]Other studies have considered, for example, standard deviations which mitigate the effect of extreme values; see Martens (2002) and Ebens (1999). However, variances appear to be the standard in the literature cited here; also see Poon and Granger (2001a).

[10]This model is interchangeably referred to as the random walk (Heynen, 1995), historical volatility (Figlewski, 1997), historical mean (Brailsford and Faff, 1996) or historical average (Yu, 2000) model.

respectively and with definitions as given in section 4.1.

We construct volatility forecasts based on the above four models for forecasting horizons of 1, 2, 5, 10 and 20 days, *i.e.* for $N = 1, 2, 5, 10$ and 20. Each of the models is initially estimated over the eight-year period ending 31 December 1995. The estimation sample which consists of 2085 daily observations is then rolled forward by one observation at a time and $1046 - N$ overlapping forecasts are constructed for the four-year out-of-sample period 1 January 1996 to 31 December 1999. Hence we estimate the model parameters on a daily basis whereas Heynen (1995) selected a monthly updating frequency[11].

To assess the forecasting performance of the different models two evaluation measures are computed with "realised" volatility defined as the sum of squared observed returns over the relevant forecasting horizon N, so $\tilde{\sigma}^2_{T+1,T+N} = \sum_{j=1}^{N} y^2_{T+j}$. The measures we use here are the regression-based R^2 statistic and the Root Mean Squared Error (RMSE) where the latter is self-explanatory. The R^2 test statistics are obtained from the OLS regression $\tilde{\sigma}^2_{T+1,T+N} = a + b\hat{\sigma}^2_{T+1,T+N} + \epsilon$. Other methods have been proposed and applied but we have chosen these as they appear to be the most commonly used[12].

4.2.2 Forecasting Results

The results of our estimation study are presented in tables 4.2 and 4.3[13]. Table 4.2 compares the forecasting performance of the four models on the basis of the R^2 statistic. These results give strong support for the use of the SV model as it almost consistently outperforms the other three models for the Standard & Poor's, Topix, DAX, CAC and AEX indices; the GARCH model usually ranks second. The main exception is the Financial Times All Share Index series for which we obtain mixed results: depending on the length of the forecasting horizon the best results are either produced by the GARCH, SV or the EWMA model. On the other hand, the simple RW model, which had proven reasonably

---

[11]Short-term volatility is a main determinant in the forecasting formulas of daily SV and GARCH models, see section 4.1, and the same applies to daily EWMA models with near-unity estimates for ϕ. Ideally, parameter estimates should therefore be updated as soon as new price information becomes available.

[12]Other evaluation methods include the Mean Squared Error (MSE), Mean Absolute Error (MAE) and variations thereof as well as the less popular Theil-U and the proportion of explained variability P statistics; see Poon and Granger (2001a) for a concise overview.

[13]Estimation results for the SV model where obtained using exact maximum likelihood methods as described in Appendices A and B. The relevant program and the data series used in this chapter can be found on the Internet at `www.ssfpack.com/holbook`; also see Appendix C.

Table 4.2. Forecasting results evaluated on the OLS regression-based R^2 statistic for six different stock indices, four different models and five forecasting horizons over the out-of-sample period 1 January 1996 to 31 December 1999

Stock Index	Model	Forecasting Horizon (in days) 1	2	5	10	20
FT All:	RW	0.006	0.010	0.016	0.020	0.024
	EWMA	0.141	0.231	0.347	0.414	**0.477**
	SV	0.146	0.250	**0.373**	**0.421**	0.454
	GARCH	**0.147**	**0.251**	0.370	0.414	0.446
S&P:	RW	0.003	0.004	0.009	0.014	0.025
	EWMA	0.029	0.045	0.084	0.105	0.110
	SV	**0.047**	**0.066**	**0.110**	**0.142**	**0.154**
	GARCH	0.038	0.055	0.096	0.120	0.124
Topix:	RW	0.013	0.022	0.037	0.054	0.078
	EWMA	0.072	0.117	0.183	0.212	0.195
	SV	0.081	**0.141**	**0.198**	**0.227**	**0.285**
	GARCH	**0.082**	0.140	0.193	0.211	0.251
DAX:	RW	0.000	0.000	0.001	0.001	0.005
	EWMA	0.086	0.131	0.197	0.235	0.259
	SV	**0.127**	**0.186**	**0.273**	**0.312**	**0.338**
	GARCH	0.115	0.163	0.233	0.258	0.267
CAC:	RW	0.005	0.008	0.013	0.017	0.023
	EWMA	0.087	0.140	0.212	0.257	0.275
	SV	**0.109**	**0.174**	**0.255**	**0.302**	**0.314**
	GARCH	0.106	0.168	0.241	0.279	0.274
AEX:	RW	0.002	0.005	0.008	0.010	0.012
	EWMA	0.123	0.190	0.261	0.280	0.265
	SV	**0.157**	**0.254**	**0.354**	**0.380**	**0.340**
	GARCH	0.149	0.240	0.334	0.360	0.316

The highest values for the goodness-of-fit R^2 statistics are in bold lettertype.

Table 4.3. Forecasting results evaluated on the basis of the Root Mean Squared Error (RMSE) statistic for six stock indices, four different models and five forecasting horizons over the out-of-sample period 1 January 1996 to 31 December 1999

Stock Index	Model	Forecasting Horizon (in days) 1	2	5	10	20
FT All:	RW	1.517	2.342	4.690	8.383	15.256
	EWMA	1.399	2.041	3.784	6.478	11.373
	SV	1.393	2.011	3.679	6.323	11.279
	GARCH	**1.386**	**1.991**	**3.617**	**6.188**	**10.869**
S&P:	RW	3.028	4.771	8.699	14.321	24.297
	EWMA	2.985	4.676	8.409	13.973	24.709
	SV	**2.915**	**4.511**	**7.883**	**12.522**	**20.778**
	GARCH	2.930	4.544	7.982	12.838	21.770
Topix:	RW	2.990	4.520	8.583	14.370	24.331
	EWMA	2.914	4.348	8.143	13.953	25.822
	SV	**2.874**	**4.211**	**7.754**	**12.794**	**21.035**
	GARCH	2.884	4.239	7.914	13.291	21.904
DAX:	RW	4.676	7.500	14.646	25.326	44.412
	EWMA	4.469	7.008	13.313	22.945	41.206
	SV	**4.340**	**6.697**	**12.311**	**20.709**	**36.105**
	GARCH	4.350	6.741	12.473	21.061	36.529
CAC:	RW	3.402	5.278	10.213	17.472	30.828
	EWMA	3.273	4.960	9.359	16.002	29.177
	SV	3.225	4.834	8.949	14.982	26.638
	GARCH	**3.217**	**4.816**	**8.906**	**14.897**	**26.577**
AEX:	RW	3.986	6.339	13.009	23.464	43.155
	EWMA	3.668	5.567	10.923	19.732	37.743
	SV	**3.582**	**5.301**	9.998	17.549	33.097
	GARCH	**3.582**	**5.301**	**9.981**	**17.440**	**32.943**

The lowest error statistic values are in bold lettertype.

successful in the study of Franses and van Dijk (1996), performs very poorly which we mainly attribute to the fact that the RW model forecasts respond very slowly when volatility changes from a low to a high state and vice versa with all observations of the previous eight years weighted equally. By comparison, the EWMA model adapts much quicker in our study due to the relatively low value we selected for the sample length of the moving average term L[14]. We further note that the highest R^2 statistics are observed for the Financial Times All Share and the AEX Index whereas the lowest are those of the Standard & Poor's series.

When we evaluate the volatility forecasts on the basis of the RMSE error statistic in table 4.3 we obtain different results. The worst performing forecasting method continues to be the RW model. Although the SV model is still superior for the Standard & Poor's, Topix and DAX indices with GARCH ranking as second best, the reverse holds with regard to the remaining three indices which at first glance appears surprising. On closer examination we found that all models had a tendency to over- rather than underpredict realised volatility with overestimation occurring approximately twice as often when we considered a one-day forecasting horizon. However, the average forecasting errors in the case of underestimation were much larger than when volatility was overestimated and squaring these forecasting errors for the calculation of the symmetric RMSE statistic penalised underestimation more heavily[15]. So, even though GARCH models almost always had the highest number of overestimations, its RMSE statistics were lowest for three of the indices[16].

We therefore conclude that the most accurate stock market volatility forecasts are produced by the SV and GARCH models and that these results are consistent across forecasting horizons. Although their ranking is sensitive to the choice of evaluation method our findings appear to give stronger support to the SV than the GARCH model and we therefore consider it the preferred volatility model for stock market volatility forecasting thus confirming the earlier results of both Heynen (1995) and Yu (2000).

[14] We also examined the EWMA model forecasting performance for a five-year moving average with L equal to 1304 and found R^2 values close to those observed for the RW model.

[15] The highest average underestimation forecast error amounted to 7.72 whereas the overestimation equivalent was 2.29.

[16] We also evaluated the forecasting power of the four models with the Mean Absolute Error (MAE) statistic (results available upon request) for which the SV and GARCH models ranked first and second again; the RW model which had the lowest number of overestimations also performed rather well, especially at the shorter forecasting horizons. Alternatively, asymmetric loss functions such as the LINEX loss function or the so-called mean mixed error statistics advocated by Brailsford and Faff (1996) could be used to provide further insight.

Chapter 5

IMPLIED VOLATILITY

In the previous chapters we have used historical data and volatility models for the calculation of volatility. An alternative source of volatility information is contained in implied volatility, which is often referred to as the market's perception of future volatility over the remaining life of the option. Historical volatility, on the other hand, is a retrospective volatility measure and, provided that the historical data is available, can be calculated for any variable whereas implied volatility is only available for those financial assets on which options are traded. Implied volatility is obtained from option prices in conjuncture with a certain option pricing model. This instantaneous volatility measure is calculated by inserting an implied volatility value in an option pricing formula so that the resulting theoretical value of the option equals its market price. The accuracy of inferred implied volatility therefore not only depends on the efficiency with which the option market subsumes the available information, but also on the use of the correct option pricing model, *i.e.* the model used by the market to price volatility.

5.1 The Black-Scholes Option Pricing Model

The classic option pricing formula of Black and Scholes (1973), who were the first to develop an explicit general equilibrium solution to the option pricing problem, states that the value of a non-dividend paying European call option on a certain asset is given by

$$c = SN(d_1) - Xe^{-r(T-t)}N(d_2) \tag{5.1}$$

with

$$d_1 = \frac{\ln(S/X) + (r + \frac{\sigma^2}{2})(T-t)}{\sigma\sqrt{T-t}},$$
$$d_2 = d_1 - \sigma\sqrt{T-t},$$

and where S is the price of the underlying asset, X is the exercise price, r is the risk free rate of return, σ is the volatility of the underlying asset which is assumed to be constant over the life of the option and $(T-t)$ is the time that remains to maturity. $N(.)$ is the cumulative probability distribution function for a standardised normal variable given by $N(d) = \frac{1}{\sqrt{2\pi}}\int_{-\infty}^{d}\exp(-\frac{x^2}{2})dx$. At time t all these values, with the exception of σ, are known and since the market price of the call option can also be observed, the implied volatility over the remaining life of the option can be calculated recursively. Unfortunately, no explicit formula for implied volatility exists as the Black-Scholes formula cannot be inverted with respect to the volatility parameter σ. It is however common to use the Newton-Raphson procedure in which case a quite accurate estimate of σ can be obtained in a relatively small number of iterations. The method starts with a reasonable volatility estimate x_0 which is revised by subtracting a value equal to the theoretical call price $c(x_0)$ minus the actual price of the call, c_m, and dividing this by the derivative of the option formula with respect to the volatility estimate, so $x_1 = x_0 - \frac{c(x_0)-c_m}{\delta c/\delta\sigma}$, where $\delta c/\delta\sigma = S\sqrt{(T-t)}\frac{1}{\sqrt{2\pi}}e^{-d_1^2/2}$, which is also known as the option's vega[1].

Implied volatility can also be inferred from put options; given the put-call parity relationship the price of a non-dividend paying European put option can be written as

$$\begin{aligned} p &= c + Xe^{-r(T-t)} - S, \\ &= S(N(d_1)-1) - Xe^{-r(T-t)}(N(d_2)-1) \\ &= -SN(-d_1) + Xe^{-r(T-t)}N(-d_2), \end{aligned} \tag{5.2}$$

with the same parameters definitions as before.

The Black-Scholes formula has however some restrictive assumptions, some of which we have already mentioned in passing above. One of these is that the underlying asset does not pay dividends (nor any other

[1] See Kritzman(1991) for a review on the application of the Newton-Raphson and the bisection method. Also see Brenner and Subrahmanyam (1988) who derive a simple implied volatility formula for the inverted Black-Scholes model which is most accurate for at-the-money options.

distributions) during the remaining life of the option. When it does, the Black-Scholes formula can be adjusted to account for this by replacing the price of the underlying asset S with $(S - PVD)$, which results in the so-called dividend-adjusted Black-Scholes formula and where PVD denotes the present value of the expected dividends that are to be paid on the underlying asset during the remaining life of the option. Alternatively, the stream of dividend payments can be assumed to be continuous over the remaining life of the option, as suggested by Merton (1973). In that case S in the Black-Scholes model can be substituted with $Se^{-\delta(T-t)}$ where δ denotes the continuous dividend yield[2].

Further, the model only applies to European options which, unlike their American counterparts, cannot be exercised prior to their maturity date. In the absence of dividend payments the prices of American and European call options are identical, so $C = c$ where C denotes the American call price and c the price of the European call. However, this equality no longer holds when the underlying asset is dividend paying as in that case early exercise of the American call option might be optimal just prior to the last ex-dividend date[3]. The dividend-adjusted Black-Scholes formula cannot therefore be applied for the pricing of dividend-paying American call options as it disregards the early exercise premium[4]. Provided that the dividend stream is fairly evenly distributed over time adapted versions of the Black-Scholes formula as proposed by Roll (1977), Geske (1979) and Whaley (1981) might be used. Unlike American calls, the price of American puts is never equal to that of its European counterpart. It was first argued by Merton (1973) that the price of an American put option, irrespective of any possible dividend payments during its lifetime, should be no lower than that of an equivalent European put as the American put option contains all features of the European put plus the right of early exercise. This additional feature should have a price, as the American put will virtually always have a positive probability of premature exercise[5]. Early exercise will become more attractive as the stock price decreases, the riskless rate increases and volatility decreases. Due to the complications arising from

[2] Also see Harvey and Whaley (1992) who show that ad hoc valuation procedures with regard to dividend payments on the Standard & Poor's 100 index can give rise to large pricing errors.

[3] Following a large dividend payment the price of the underlying asset, and hence the price of the option, may decrease substantially. Provided that the option is sufficiently in-the-money it might be beneficial to sacrifice the remaining time value of the option.

[4] See Whaley (1982) who empirically compares several methods for the pricing of American call options on dividend-paying stocks.

[5] For example, when the price of the underlying assets falls far below the exercise price, the put might prove to be more valuable when exercised immediately as the proceeds can be invested in bonds which will earn a riskless rate.

a possible early exercise the put-call parity theorem does not hold for American options which implies distinct treatment of call and put options. An alternative is to use the binomial tree approach of Cox, Ross and Rubinstein (1979) which can be extended to value American style options on dividend paying assets.

One of the main assumptions underlying the Black-Scholes option pricing formula concerns the price behaviour of the underlying asset which is assumed to follow a random walk in continuous time with a standard deviation rate proportional to the square root of time. Assuming lognormality, the log price follows a normal probability distribution such that the mean and standard deviation of $\ln(S_T)$ are shown to be

$$\ln(S_T) \sim \mathrm{N}(\ln(S_t) + (\mu - \frac{\sigma^2}{2})(T-t), \sigma\sqrt{T-t}),$$

and

$$\ln(\frac{S_T}{S_t}) \sim \mathrm{N}((\mu - \frac{\sigma^2}{2})(T-t), \sigma\sqrt{T-t}),$$

where S_T is the future price of the asset at time T, S_t its current price, μ is the expected return of the proportional change in the underlying asset and σ its volatility. It is apparent from the above that, given any fixed length for the time interval, the volatility of the continuously compounded returns is assumed to be constant. Substantial empirical research has however unambiguously shown that volatility is time-varying[6].

If the Black-Scholes formula would be an appropriate formula for the description of option prices, the resulting values for implied volatility would have to be identical across options which only differ with respect to their strike or exercise price. The strike structure of volatility is however found to depend on the moneyness of the options and frequently takes the form of an U-shaped curve, also known as the volatility "smile" or "smirk"[7]. The further options are in or out-of-the-money, the higher their implied volatility; the lowest values for implied volatility are then observed for at-the-money options. Similarly, implied volatilities can vary by their time to maturity, and are known to do so, even though the assumption of constant volatility implies a flat term structure[8]. The term structure of volatility slopes upward when short-term implied

[6] An option pricing model which allows for stochastic volatility was introduced by Hull and White (1987).
[7] See, for example, Rubinstein (1985), Stein (1989), Stein and Stein (1991) and Canina and Figlewski (1993).
[8] See, for example, Stein (1989) and Xu and Taylor (1994).

volatilities are low, and is downward sloping when short-term implied volatilities are high[9].

The above could interpreted as evidence that the Black-Scholes model is an unsatisfactory option pricing model as the calculated values for implied volatility are found to be biased. These biases can however also be (partially) explained by measurement errors relating to the other variables in the model. Measurement errors can arise from a number of sources, such as the non-simultaneous observation of option prices and the price of the underlying asset, bid-ask spreads, discrete tick prices and the fact that the Black-Scholes formula is non-linear. Liquidity of the option considerably mitigates the problems relating to non-simultaneity and the bid-ask spread where it should be noted that, given the underlying asset of the option, at-the-money options are often the most frequently traded ones. Highly liquid option series do not only have smaller bid-ask spreads but it is also increasingly likely that the last trading prices of the day were observed just prior to the closure of the option market; provided that the option market and the market on which the underlying asset is traded close at the same time, frequent trading therefore alleviates the problem of non-simultaneity. With regard to the non-linearity of the Black-Scholes formula, it has been shown by Feinstein (1995) that the Black-Scholes model gives virtually unbiased implied volatility estimates for at-the-money options. For the calculation of implied volatility these options are therefore to be preferred to options that are further into or out-of-the-money.

5.2 Forecasting with Implied Volatility: Empirical evidence

The information value of implied volatility has been established in numerous empirical studies. The hypothesis tested is that the information content of implied volatility should subsume that of all other variables in the information set, provided that the option market is efficient and that the correct option model has been applied to infer the implied volatility measure. In that case implied volatility forecasts should be more accurate than historical volatility forecasts.

Since option series do not result in unique values for implied volatility the question arises which option series should be selected in order to obtain the market's volatility forecast with regard to the underlying

[9]See Whaley (1993) who observes a downward sloping term structure for OEX options with a dramatic increase in implied volatility during their final week of trading. As a possible explanation he suggests that this might be attributable to increased speculation when options are close to maturity.

asset[10]. Following initial research by Latané and Rendleman (1976) implied volatility is often inferred from a number of options and a weighting scheme is applied. The simplest approach is to weight all implied volatilities equally as in, for example, Schmalensee and Trippi (1978). Above it was shown however than certain options result in more accurate implied volatility measures than others and it therefore appears reasonable to give more weight to such options. Schmalensee and Trippi (1978) circumvented this problem by simply excluding those options that were far into or out-of-the-money and options which were close to maturity. Latané and Rendleman (1976) related their weights to the option's vega where the options that exhibited a higher degree of sensitivity to the implied volatility measure were weighted more heavily, thus emphasising implied volatility inferred from at-the-money options. They examined 24 individual stocks using weekly observations from October 1973 to June 1974. Although Latané and Rendleman (1976) ignored dividend payments they found that their implied volatility measure was highly correlated with subsequently realised volatility and outperformed volatility predictors calculated from historical data. Their results were confirmed by Chiras and Manaster (1978) who did account for dividends by converting realised dividends into a continuous dividend yield and using Merton's (1973) dividend adjusted formula for monthly observations of Chicago Board Options Exchange (CBOE) stock options between June 1973 and April 1975. They obtained implied volatility measures by applying a weighting scheme which assigned weights to the various implied volatility measures on the basis of the option's price elasticity with respect to its implied standard deviation. It was shown by Day and Lewis (1988) that the latter method assigns more weight to out-of-the-money options than the weighting scheme based on vegas. Empirically, the forecasting accuracy of the different weighting techniques was addressed by Beckers (1981) who compared various techniques for daily stocks options between October 1975 and January 1976. Apart from the observation that implied volatility yielded more accurate forecasts of future volatility than historical standard deviations, his study indicated that an alternative technique which effectively minimised the weighted sum of the squared deviations between the market price and the corresponding Black-Scholes price of the option tended to outperform the method of Latané and Rendleman (1976)[11]. Both methods were however inferior

[10]The rationale behind using more than one option series is that the noise present in implied volatilities is diversified.

[11]Also see Whaley (1982) who obtained similar results using a variety of option pricing models, which resulted in approximately the same implied volatility values.

predictors of realised volatility when compared with an implied volatility measure which consisted of the implied volatility for the most sensitive option, which was the at-the-money option[12]. Feinstein (1989) reported findings along the same lines rejecting longer and medium-term forecasts based on historical standard deviations and a number of weighting schemes in favour of methods which were exclusively constructed from implied volatilities of just-out-of-the-money call options. The early literature therefore supported the hypothesis that implied volatility provided a more accurate forecast of future volatility than historical volatility and that the most accurate implied volatility forecasts were those obtained from those options that were at or near-the-money. Surprisingly, the opposite conclusion was reached by Canina and Figlewski (1993) who found implied volatility to be virtually uncorrelated with subsequent realised volatility and favoured a simple historical volatility measure. In their study they examined the information value of daily Standard & Poor's 100 stock index call options over the period May 1973 to May 1987. Implied volatility values were obtained using a 500-step binomial method and grouped on the basis of time-to-maturity and intrinsic value. Both the thus obtained implied volatilities and a simple 60-day historical volatility measure were regressed on realised volatility for evaluation purposes. Although the regression coefficients were statically insignificant for the implied volatilities, historical volatility appeared to have some explanatory power which was confirmed in an encompassing regression[13].

An alternative approach to historical volatility forecasts are time-varying volatility models, such as the Generalised Autoregressive Conditional Heteroskedasticity (GARCH) models, and an increasing number of the later studies formulated historically based volatility forecasts in terms of these type of models. The first comparative study in which GARCH models were employed was conducted by Day and Lewis (1992) who developed GARCH models which incorporated implied volatility as an explanatory variable in the variance equation, thus obtaining nested models which facilitated the use of likelihood ratio tests. They extracted implied volatilities from call options on the Standard & Poor's 100 Index attaching weights on the basis of trading volume and examined its informational content at a weekly frequency over the period March 1983 to December 1989. Although their results indicated that GARCH models

[12]Although the most sensitive option is usually slightly out-of-the-money Beckers (1981) prefers to refer to this option in his paper as being "at-the-money".

[13]The possible explanations offered by Canina and Figlewski (1993) for their findings are discussed by Christensen and Prabhala (1998).

provided better volatility forecasts than implied volatility, they found that the best forecasts were obtained when both sources of information were used. Lamoureux and Lastrapes (1993) obtained similar results for daily options on ten individual actively traded stocks over the period April 1982 to March 1984 where implied volatility was inferred using the stochastic volatility model of Hull and White (1987). Examining both in-sample and out-of-sample forecasting abilities they concluded that they were unable to reject the hypothesis that historical data contained incremental information.

In the foreign exchange literature a different picture emerged as studies by Jorion (1995), Xu and Taylor (1995) and Amin and Ng (1997) showed that the volatility information contained in GARCH models is subsumed by the implied volatility measures. The assumed comprehensive information content of implied volatility for stock indices is however supported in a number of later studies. Christensen and Prabhala (1998) find evidence that implied volatility outperforms historical volatility for options on the Standard & Poor's 100 stock index which are examined at a monthly frequency over the longer period November 1983 to May 1995. In addition they observe that for some of their specifications the OEX implied volatilities subsume the information content of past volatility and that implied volatility has become a better predictor of future volatility following the 1987 stock market crash. The same stock index options are examined by Fleming (1998) who extracts daily OEX implied volatilities between October 1985 and April 1992 using a modified binomial model. Although he finds implied volatility to be an upwardly biased volatility predictor it outperforms the historically based methods in terms of ex ante forecasting ability. His findings are confirmed by those of Blair, Poon and Taylor (2001) who examine a GARCH model with embedded implied volatility and high frequency returns using the VIX index of the CBOE as a measure of daily implied volatility on OEX options. Not only do they observe that the most accurate volatility forecasts are obtained with implied volatility, they also find that there is little or no incremental information about future volatility contained either in the GARCH models or in the intraday returns.

Chapter 6

FORECASTING THE VARIABILITY OF STOCK INDEX RETURNS WITH STOCHASTIC VOLATILITY MODELS AND IMPLIED VOLATILITY

6.1 Introduction

Forecasts of financial market volatility play a crucial role in financial decision making and the need for accurate forecasts is apparent in a number of areas. All investors face the decision whether or not to hedge the risks associated with their investments. Portfolio diversification strategies can be applied to reduce the total risk of a portfolio but exposures can be even further reduced by means of hedging strategies. Investors will then base their hedging decisions on their risk perception over the remaining investment horizon; the more volatile the market the more inclined investors will be to hedge their exposures. Instruments that are commonly used for this purpose are financial derivatives such as options and futures. The pricing of these sometimes very intricate instruments largely depends on the risk associated with the underlying asset and volatility is therefore also a key input parameter is many derivatives pricing models. The issue of accurate volatility forecasts is therefore firmly positioned at the centre of financial decision making. Unfortunately, it is notoriously difficult to predict volatility accurately and the problem is exacerbated by the fact that realised volatility has to be approximated as it is inherently unobservable. Due to its critical role, the topic of volatility forecasting has however received much attention and the resulting literature is considerable[1].

One of the main sources of volatility forecasts are historical parametric volatility models such as Generalised Autoregressive Conditional Het-

[1] A comprehensive survey of the findings in the volatility forecasting performance literature is given in Poon and Granger (2001a, 2001b).

eroskedasticity (GARCH) and Stochastic Volatility (SV) models. The parameters in these models are estimated with historical data and subsequently used to construct out-of-sample volatility forecasts. The high degree of intertemporal volatility persistence observed by these models suggests that the variability of stock index returns is highly predictable and that past observations contain valuable information for the prediction of future volatility. Studies comparing the forecasting abilities of the various volatility models have been undertaken for a number of stock indices and the general consensus appears to be that those models that attribute more weight to recent observations outperform others[2].

An alternative information source for volatility prediction is found in implied volatility and as this volatility measure is forward looking its information content should, under certain conditions, subsume that of all other variables in the information set.

The question whether the most accurate volatility forecasts are in fact produced by implied volatility, rather than by the historically based volatility models, has been researched extensively in the GARCH literature and the most recent of these studies appear to confirm the superiority of implied volatility as a source for future volatility information. Thus far the issue of comparative forecasting ability has however not been studied in the context of SV models. In the past few years this class of volatility model has received considerable attention in the literature and it can be regarded as a competitive alternative to GARCH models even though its empirical application has been limited. In this chapter we examine the predictive ability of the SV model and compare its volatility forecasts with those of implied volatility. For this purpose we introduce an SV model which incorporates implied volatility as an exogenous variable in the variance equation. This model, which we will refer to as the Stochastic Volatility with eXogenous variables (SVX) model, allows us to perform statistical tests for nested models. We evaluate the predictive performance for daily returns on the Standard & Poor's 100 index and as a measure of implied volatility we use the VIX index of the Chicago Board Options Exchange (CBOE). In addition, we compare the ex-ante forecasting ability of the different methods over a four-and-a-half year evaluation period for forecasting horizons ranging from 1 to 20 trading days. As measures of realised volatility we consider both daily squared returns and an intraday volatility measure which is calculated from intraday squared returns.

[2] See *e.g.* Akgiray (1989), Dimson and Marsh (1990) and Walsh and Tsou (1998) for an overview.

The remainder of this chapter is structured as follows. In the next section we discuss the various model specifications. The data and in-sample estimation results are presented in section 6.3. Section 6.4 provides details of our forecasting methodology and the out-of-sample forecasting results are given in section 6.5. In the final section we conclude and provide a summary.

6.2 Model Specifications

Generalised Autoregressive Conditional Heteroskedasticity (GARCH) models have thus far been the most frequently applied class of time-varying volatility model. Since its introduction by Engle (1982) and subsequent generalisation by Bollerslev (1986) this model has been extended in numerous ways which usually involved alternative formulations for the volatility process[3]. Although the Stochastic Volatility (SV) model has been recognised as a viable alternative to the GARCH model the latter is still the standard in empirical applications[4]. This is mainly due to the problems which arise as a consequence of the intractability of the likelihood function of the SV model which prohibits its direct evaluation. However, in recent years considerable progress has been made in this area which does not only encourage further empirical research but also enables the development of various extensions of the SV model. One of the possible extensions involves the inclusion of exogenous variables in the variance equation which we will discuss in this chapter; the resulting model we refer to as the SVX model.

Volatility models are usually defined by their first two moments, the mean and the variance equation. The general notation for the mean equation of time-varying volatility models is given by

$$y_t = \mu_t + \sigma_t \varepsilon_t, \qquad \varepsilon_t \sim \text{NID}(0,1), \quad t = 1, \ldots, T, \tag{6.1}$$

where y_t denotes the return series of interest and μ_t its conditional mean[5]. The disturbance term ε_t is assumed to be identically and in-

[3]For surveys on GARCH models we refer to Bollerslev, Chou and Kroner (1992), Bera and Higgins (1993), Bollerslev, Engle and Nelson (1994) and Diebold and Lopez (1995).

[4]SV models are reviewed in, for example, Taylor (1994), Ghysels, Harvey and Renault (1996) and Shephard (1996).

[5]For SV models the conditional mean is usually assumed to be equal to zero or is modelled prior to estimation of the volatility process. Simultaneous estimation of the mean and variance equation has been undertaken in, for example, our study of the Stochastic Volatility in Mean model in Chapter 3.

dependently distributed with zero mean and unit variance. In addition, the assumption of normality is added. A common notation for the variance equation of the SV class of volatility models is given by

$$\sigma_t^2 = \sigma^{*2} \exp(h_t), \tag{6.2}$$

and it is therefore defined as the product of a positive scaling factor σ^{*2} and the exponential of the stochastic process h_t. For the standard SV model h_t is specified as a first-order autoregressive process in order to capture the well-documented volatility clustering phenomenon, so

$$h_t = \phi h_{t-1} + \sigma_\eta \eta_t, \qquad \eta_t \sim \text{NID}(0,1), \tag{6.3}$$

where the degree of volatility persistence is measured by the ϕ parameter which is restricted to a positive value smaller than one in order to ensure the stationarity of the volatility process, so $0 < \phi < 1$. Further, it is assumed that the disturbance term η_t is mutually uncorrelated with the error term ε_t in the mean equation (6.1), both contemporaneously and at all lags. The SV model with embedded implied volatility is labelled the SVX model and we could specify its stochastic process as

$$h_t = \phi h_{t-1} + \gamma x_{t-1} + \sigma_\eta \eta_t, \qquad \eta_t \sim \text{NID}(0,1), \tag{6.4}$$

where x_{t-1} denotes the lagged implied volatility measure in logarithmic squared form, so $x_{t-1} = \ln \sigma_{IV,t-1}^2$. The value for ϕ in the SVX model is restricted to be less than one in absolute values, *i.e.* $-1 < \phi < 1$. The equivalent notation is commonly used in the GARCH literature even though it implicitly includes an entire lag structure of the implied volatility measure. For the SVX model this becomes apparent when we rewrite the volatility process in logarithmic terms as

$$\begin{aligned} \ln \sigma_t^2 &= \ln \sigma^{*2} + h_t \\ &= \ln \sigma^{*2} + \phi h_{t-1} + \gamma x_{t-1} + \sigma_\eta \eta_t \\ &= (1-\phi) \ln \sigma^{*2} + \phi \ln \sigma_{t-1}^2 + \gamma x_{t-1} + \sigma_\eta \eta_t, \end{aligned}$$

and if we repeatedly substitute for the lagged volatility process we observe that for large values of t

$$\ln \sigma_t^2 = \ln \sigma^{*2} + \gamma x_{t-1} + \gamma \sum_{i=2}^{t-1} \phi^{i-1} x_{t-i} + \sigma_\eta \sum_{i=0}^{t-1} \phi^i \eta_{t-i}.$$

In comparison, the equivalent notation for the SV model with h_t as defined in equation (6.3) can be written as

$$\ln \sigma_t^2 = \ln \sigma^{*2} + \sigma_\eta \sum_{i=0}^{t-1} \phi^i \eta_{t-i}.$$

Specification of the h_t process as in equation (6.4) therefore imposes an exponential lag distribution for the implied volatility measure which leads to a downwardly biased value for $\hat{\gamma}$ when $\hat{\phi}$ is positive. For a negative $\hat{\phi}$ parameter the value for $\hat{\gamma}$ will be asymptotically upwardly biased, as $\sum_{i=1}^{\infty} \phi^i < 0$ for $-1 < \phi < 0$. Obviously, the size of this bias depends on the estimated value for the persistence parameter ϕ. If $\hat{\phi}$ is close to zero and insignificant, *i.e.* if all volatility information is impounded in the implied volatility measure, $\hat{\gamma}$ will only be marginally biased. For the GARCH class of models the issue of a distributed lagged implied volatility structure was pointed out by Amin and Ng (1997) who suggested a persistence adjustment term[6]. A similar structure can be implemented for the SVX model by defining h_t as follows

$$h_t = \phi h_{t-1} + \gamma(1 - \phi L) x_{t-1} + \sigma_\eta \eta_t, \qquad \eta_t \sim \text{NID}(0, 1), \tag{6.5}$$

which by recursive substitution of the logarithmic variance equation leads to

$$\begin{aligned} \ln \sigma_t^2 &= (1 - \phi) \ln \sigma^{*2} + \phi \ln \sigma_{t-1}^2 + \gamma(1 - \phi L) x_{t-1} + \sigma_\eta \eta_t, \\ &= \ln \sigma^{*2} + \gamma x_{t-1} + \sigma_\eta \sum_{i=0}^{t-1} \phi^i \eta_{t-i}, \end{aligned}$$

and therefore omits the implied volatility lag structure. By defining h_t as in equation (6.5) we therefore obtain an alternative SVX model, which we will refer to hereafter as the SVX model with persistence adjustment, or the SVX$^+$ model. When implied volatility $\sigma_{IV,t}$ is regarded as an

[6] A multiplicative error correction had previously been used by Baillie and Bollerslev (1989) in a GARCH model with seasonal dummies.

efficient predictor of volatility σ_t, then $\ln \sigma_{IV,t}^2$ is a biased estimator for $\ln \sigma_t^2$. However, the equation for $\ln \sigma_t^2$ includes the constant term $(1-\phi)\ln \sigma^{*2}$ which accounts for this bias.

Finally, we consider a restricted volatility model by imposing $\phi = 0$ on equation (6.4) or (6.5) and the h_t definition of this model is therefore given by

$$h_t = \gamma x_{t-1} + \sigma_\eta \eta_t, \tag{6.6}$$

from which it follows that

$$\ln \sigma_t^2 = \ln \sigma^{*2} + \gamma x_{t-1} + \sigma_\eta \eta_t, \qquad \eta_t \sim \text{NID}(0,1).$$

This last of our four models we term the Stochastic Implied Volatility (SIV) model as the volatility process is determined by the implied volatility measure and a stochastic error term η_t.

Various estimation methods have been proposed in the literature for the basic SV model. A quasi-maximum likelihood method based on a linearised representation of the SV model was developed by Harvey, Ruiz and Shephard (1994) while Sandmann and Koopman (1998) used importance sampling methods to obtain exact maximum likelihood estimates for the same linearised model. Danielsson (1994) introduced the use of importance sampling techniques for the estimation of SV models. Other methods of estimating SV models are discussed in Kim, Shephard and Chib (1998), Fridman and Harris (1998) and Watanabe (1999). Bayesian approaches to estimating SV models have been explored by Jacquier, Polson and Rossi (1994) and Shephard and Pitt (1997). An excellent review of stochastic volatility, its relevance to financial economic theory, SV models and estimation is given by Ghysels, Harvey and Renault (1996). Here we adopt importance sampling methods developed for non-Gaussian and non-linear state space models by Shephard and Pitt (1997) and Durbin and Koopman (1997, 2000). The SV and SVX classes of models can be considered as special cases of non-linear state space models. The parameters of the SVX class of models are estimated by simulated maximum likelihood using importance sampling techniques which are described in Appendix B.

6.3 Data Description and Empirical In-Sample Results

6.3.1 Data

Daily stock index returns

In our empirical study we analyse the Standard & Poor's 100 daily stock index series for the period 2 January 1986 to 29 June 2001. The data are obtained from Datastream and after adjusting the series for holidays our sample consists of 3906 daily observations. The continuously compounded returns on the stock index are expressed in percentage terms and are therefore given by $R_t = 100(\ln P_t - \ln P_{t-1})$ where P_t denotes the closing price of the Standard & Poor's 100 index at time t.

Implied volatility measures

Implied volatility is often referred to as the market's perception of future volatility of the underlying asset over the remaining life of the option. This measure of volatility is calculated in conjuncture with an option pricing model and the remaining input variables in the model, such as the option price, the price of the underlying asset, the exercise price, time to maturity, future dividend payments and their timing, the risk free rate of return and any other provisions contained in the option contract, *e.g.* American options can be exercised prior to maturity. Even when the correct option pricing model, *i.e.* the model used by the market to price volatility, has been used the resulting implied volatility measure can still be biased due to measurement errors in the model input variables. Such errors can for example be caused by non-simultaneous observation of the option price and the price of the underlying asset, bid-ask spreads and discrete time ticks.

The implied volatility index we use in our study is the Chicago Board Options Exchange Market Volatility Index (VIX) which is based on a highly liquid options market. The VIX data was extracted from the CBOE on-line database for the period 2 January 1986 to 29 June 2001[7]. Implied volatilities used for the construction of the VIX index are calculated from the midpoint of bid-ask option prices using a binomial method as described in Harvey and Whaley (1992) which takes into account the level and timing of dividend payments on the underlying Standard & Poor's 100 stock index (OEX) and the fact that OEX options are American style. The Black-Scholes model assumption of constant volatility however introduces bias into the implied volatility measure but the mag-

[7] `www.cboe.com/MktData/vix.asp#vix`

nitude of the resulting bias is smaller for options that are at-the-money and close to expiry[8]. With this in mind the implied volatility index is based on the implied volatilities of eight near-the-money, nearby and second nearby expiry call and put options where the shortest time to maturity is at least eight days in order to eliminate the increase in implied volatility observed during the option's last week of trading. Taking both call and put options has several advantages: it reduces possible bias due to staleness in the observed stock index level and increases the amount of information. The calendar-day implied volatilities of each of the eight options are adjusted to a trading base by multiplying the calendar-day rate by $\sqrt{N_t}/\sqrt{N_c}$ where N_t and N_c represent the number of trading and calendar days to expiry, respectively, which might induce an upward bias[9]. The VIX index is then constructed as a weighted average of these implied volatilities in such a way that it represents the implied volatility of a hypothetical at-the-money OEX option with twenty-two trading days to expiry[10]. Although the VIX represents an implied volatility measure for the Standard & Poor's 100 stock index which is biased, we still consider it a useful proxy of future volatility for the practical application of volatility forecasting as it mitigates many of the measurement errors which typically contribute to biased implied volatility measures.

For the calculation of our implied volatility measure we use the daily closing level of the VIX index and from the annualised VIX, which is expressed in terms of standard deviations, we calculate the daily VIX variance at time t as $\sigma^2_{IV,t} = VIX^2_t/252$.

Data Summary

In figure 6.1 we present the daily return and the VIX series, together with the squared return series which can be regarded as a noisy approximation of realised volatility[11]. In addition we show the histogram of the daily returns on the Standard & Poor's 100 index in the top-right corner which shows that this series is negatively skewed and exhibits leptokur-

[8]Hull and White (1987) found that when volatility was stochastic biases in the Black-Scholes model were smallest for near-the-money and close to maturity options. Also see Feinstein (1995).

[9]See: Fleming, Ostdiek and Whaley (1995) and Blair *et al.* (2001). Also, it was pointed out to us that ignoring the wildcard option implicit in the OEX options, see Fleming and Whaley (1994) and Whaley (1993), and the fact that index option prices may include a volatility risk premium, see Bakshi and Kapadia (2001), could further contribute to an upwardly biased implied volatility measure.

[10]The construction of the VIX index is described in more detail by Whaley (1993) and by Fleming, Ostdiek and Whaley (1995), who regard it as a useful proxy for expected stock market volatility.

[11]Note that the graph of the squared return series is truncated at a value of 100 which only affects the observation of 19 October 1987 that has a value of 561.214.

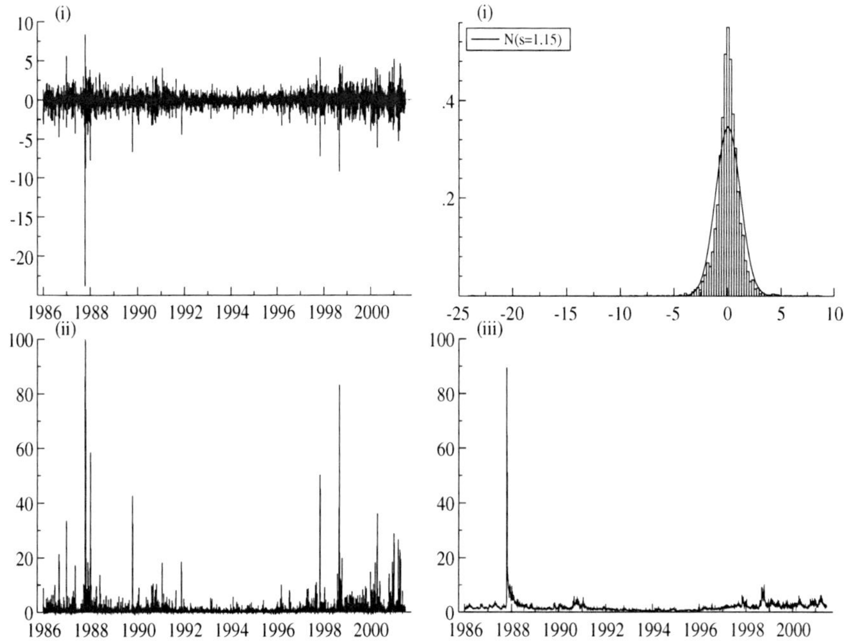

Figure 6.1. Daily (i) returns and (ii) squared returns (truncated at 100) on the Standard & Poor's 100 index and (iii) the VIX index between 02/01/86 and 29/06/01

tosis. As our full sample includes observations relating to the October 1987 stock market crash which might have a distorting influence on the estimation results we also consider a subsample that starts on 4 January 1988. Summary statistics for both samples are given in table 6.1. The effects of the large outliers in the full sample are illustrated by the very high values for the skewness and excess kurtosis coefficients. When we compare the variances of the two return series we observe a value of 1.323 for the full sample against 1.067 for the shorter sample which represent annual standard deviation values of 18.3% and 16.4%, respectively. The decrease in volatility is also reflected in the VIX series which has however much larger mean values than the squared return series. When these VIX values are translated into annual standard deviations these amount to 22.2% for the full sample and 21.1% for the post-crash period which suggests that the implied volatility measure is upwardly biased. The graphs in figure 6.1 show that the VIX and the squared return series follow a very similar pattern although the VIX series is much smoother, *i.e.* less volatile, which is confirmed by their respective

Table 6.1. Summary statistics of daily returns and squared returns on the S&P 100 Index and the VIX index

Period	1986-2001			1988-2001		
No. of Obs. T	3906			3401		
Series	S&P100		VIX	S&P100		VIX
	R_t	R_t^2	$\sigma_{IV,t}^2$	R_t	R_t^2	$\sigma_{IV,t}^2$
Mean	0.046	1.325	1.954	0.049	1.069	1.774
Variance	1.323	93.042	7.260	1.067	9.035	1.395
Skewness	-2.513	50.699	18.764	-0.428	12.585	1.802
Excess Kurtosis	51.584	2908.65	503.698	6.008	250.746	5.280
Maximum	8.539	561.214	89.521	5.606	83.258	9.668
Minimum	-23.690	0.000	0.324	-9.125	0.000	0.324
$\hat{\rho}_1$	-0.019	0.166	0.792	-0.017	0.226	0.960
$\hat{\rho}_2$	-0.066	0.136	0.630	-0.045	0.090	0.932
$\hat{\rho}_3$	-0.035	0.070	0.628	-0.040	0.055	0.914
$\hat{\rho}_4$	-0.023	0.026	0.605	0.001	0.095	0.900
$\hat{\rho}_5$	0.021	0.128	0.574	-0.025	0.142	0.876
$Q(12)$	37.432	283.50	12,238	42.945	456.57	30,672

$\hat{\rho}_\ell$ is the sample autocorrelation coefficient at lag ℓ with asymptotic standard error $1/\sqrt{T}$ and $Q(\ell)$ is the Box-Ljung portmanteau statistic based on ℓ squared autocorrelations. The critical value at the 1% significance level for the Q(12) statistic is 26.22.

variance statistics in table 6.1. Further, the degree of autocorrelation is much higher for the VIX series than for the squared return series, especially for the 1988–2001 sample where autocorrelation coefficients for the VIX series are comparable with the persistence parameter values typically found for SV and GARCH models. The Q(12) statistics for the return series indicates that the null hypothesis of zero values for the first twelve autocorrelation coefficients has to be rejected at the 1% level for both samples. The first-order autocorrelation coefficients are however not significantly different from zero even though this is frequently observed for stock index series[12]. We therefore leave the conditional mean μ_t unmodelled, *i.e.* we assume that in equation (6.1) $\mu_t = 0$.

[12]See, *e.g.*: Campbell, Lo and MacKinlay (1997, Chapter 2).

6.3.2 Empirical In-Sample Results

In this section we discuss the empirical results which were obtained by estimating the models introduced in section 6.2: the SV, the SVX, the SVX^+ and the SIV model[13]. The general mean and variance equations are defined in equations (6.1), with $y_t = R_t$ and $\mu_t = 0$, and (6.2), respectively. The log processes for these models are given by

SV Model: $$h_t = \phi h_{t-1} + \sigma_\eta \eta_t$$

SVX Model: $$h_t = \phi h_{t-1} + \gamma x_{t-1} + \sigma_\eta \eta_t$$

SVX^+ Model: $$h_t = \phi h_{t-1} + \gamma(1 - \phi L)x_{t-1} + \sigma_\eta \eta_t$$

SIV Model: $$h_t = \gamma x_{t-1} + \sigma_\eta \eta_t.$$

The SV model can be obtained by imposing the parameter restriction $\gamma = 0$ in the h_t definition of either the SVX or the SVX^+ model and when we impose $\phi = 0$ for the SVX or the SVX^+ model we obtain the SIV model. Statistical tests can therefore be based on so-called nested models.

In table 6.2 we report parameter estimates and tests for the informational content of the VIX index relative to the SV model for daily returns on the Standard & Poor's 100 index over the periods 1986–2001 and 1988–2001. For the SV model we find that the estimated coefficients for the persistence parameter ϕ are close to unity for both samples with values of 0.977 for the full sample and 0.984 for the sample starting in 1988. The fact that the shorter 1988–2001 sample does not contain the large outlier of the longer sample is reflected not only in the size of the ϕ estimate, but also in the estimated values for the scaling parameter σ^{*2} and the variance of η_t, as these are both larger for the more volatile 1986–2001 sample. The resulting unconditional variance values calculated from

$$\sigma^{*2} \exp\left(0.5 \frac{\sigma_\eta^2}{1 - \phi^2}\right),$$

also reflect the difference in the average annualised volatility in terms of standard deviations which amount to 17.0% for the full sample and 16.5% for the sample starting after the 1987 stock market crash. Of

[13] These models were estimated using programs and data series which can be found on the Internet at `www.ssfpack.com/holbook`; also see Appendix C.

Table 6.2. In-sample estimation results for returns on the Standard & Poor's 100 stock index over the period 2 January 1986 to 29 June 2001 and 4 January 1988 to 29 June 2001

Period	1986–2001				1988–2001			
T	3906				3401			
Model	SV	SVX	SVX$^+$	SIV	SV	SVX	SVX$^+$	SIV
σ^{*2}	0.804	0.431	0.432	0.432	0.780	0.430	0.429	0.431
	0.627 1.031	0.403 0.462	0.403 0.463	0.403 0.462	0.573 1.061	0.400 0.461	0.400 0.459	0.401 0.462
ϕ	0.977	−0.050	0.013		0.984	−0.133	−0.128	
	0.965 0.985	−0.238 0.142	−0.190 0.215		0.972 0.991	−0.338 0.090	−0.351 0.117	
γ		1.097	1.043	1.044		1.205	1.066	1.063
		1.025 1.168	0.974 1.112	0.975 1.112		1.120 1.290	0.991 1.142	0.987 1.139
σ_η^2	0.032	0.417	0.413	0.415	0.021	0.363	0.363	0.363
	0.022 0.046	0.346 0.504	0.340 0.503	0.344 0.501	0.013 0.034	0.288 0.459	0.287 0.460	0.287 0.457
$\ln L$	-5,310.36	-5,207.17	-5,207.30	-5,207.31	-4,491.39	-4,405.13	-4,405.30	-4,405.84
$LR(\gamma = 0)$		206.38	206.12			172.52	172.18	
$LR(\phi = 0)$		0.28	0.02			1.42	1.08	
AIC	10,626.72	10,422.34	10,422.60	10,420.62	8,988.78	8,818.26	8,818.60	8,817.68
$Q(12)$	22.866	19.555	19.744	19.700	23.951	23.578	23.482	23.787
N	16.731	43.335	43.167	43.340	14.555	32.350	32.254	32.004

Parameter estimates are reported together with the asymptotic 95% confidence intervals which are asymmetric for σ^{*2}, ϕ and σ_η^2; $LR(\gamma = 0)$ and $LR(\phi = 0)$ are the likelihood ratio statistics for the hypotheses $\gamma = 0$ and $\phi = 0$, respectively. AIC is the Akaike Information Criterion which is calculated as -2(ln L) + 2p; $Q(\ell)$ is the Box-Ljung portmanteau statistic for the estimated observation errors which is asymptotically χ^2 distributed with $\ell - p$ degrees of freedom where p is the total number of estimated parameters; N is the χ^2 normality test statistic with 2 degrees of freedom.

further interest are the diagnostic test-statistics for the estimated series of ε_t, the error term in the mean equation. They indicate that the assumption of zero serial correlation is not violated by the SV model and that the SV model is capable of absorbing the excess kurtosis found in the underlying series.

The SVX and SVX$^+$ models combine two sources of volatility information and this clearly leads to a better fit of our data-set. We find that the γ parameters in these models are always statistically significant which confirms the earlier findings in the GARCH literature that implied volatility contains crucial information about the volatility process. The estimates of the ϕ parameters in the SVX models are negative albeit insignificant; this implies that the γ parameter is upwardly biased. Inclusion of the persistence adjustment term in the SVX$^+$ model reduces the value of γ and the null hypothesis of $\gamma = 1$ can no longer be rejected for either sample. The estimated values of the first-order autoregressive term ϕ remain insignificant but are now closer to zero compared to corresponding values of the SVX model. The increase of the estimated variance σ_η^2, when compared to the SV model, is due to the inclusion of $\ln \sigma_{IV,t}^2$ as the implied volatility measure introduces additional noise to the h_t process. The substantially reduced estimated values for σ^{*2} then suggest higher levels of h_t which might further attribute to the increase in the variance of η_t. The likelihood ratio test statistics show that there is little difference between the SVX and the SVX$^+$ model. The null hypothesis $H_0 : \gamma = 0$ can never be rejected. We conclude that implied volatility is a statistically significant factor in identifying the underlying volatility.

The SIV model, that is the SVX or SVX$^+$ model with $\phi = 0$, is also estimated and we obtain similar results as for the SVX and SVX$^+$ models. This is not surprising since ϕ was not significant at the 5% significance level for these models. Also, the likelihood ratio test for $H_0 : \phi = 0$ never exceeded the critical χ^2 5% significance value of 3.84. The estimated values for γ are closest to those of the SVX$^+$ model as this model excludes the multiple lagged implied volatility structure by deducting $\phi\gamma x_{t-2}$ from the h_t definition. These results show that the SV class models which include implied volatility measures such as the VIX produce more effective estimates of the underlying volatility process of Standard & Poor's 100 stock index returns than dynamic models which are solely based on historical returns such as the standard SV model.

6.4 Volatility Forecasting Methodology

In the next section we present forecasts based on the rolling window principle. Initially we estimate parameters over the period 4 January 1988 to 3 January 1997 and which therefore starts after the stock market crash. This sample spans a period of 9 years and consists of 2273 observations. This leaves an evaluation period of 1128 observations covering four-and-a-half years of data, *i.e.* from 6 January 1997 to 29 June 2001. Having calculated the volatility forecasts based on the parameters of this initial sample we roll it forward by one trading day, thus keeping the sample size constant at 2273 observations. We also construct volatility forecasts for 2, 5, 10, 15 and 20-day horizons every time we estimate the model parameters. We therefore obtain overlapping forecasts with sample size $1129 - N$ for $N = 1, 2, 5, 10, 15$ and 20. In the remainder of this section we define the N-day ahead volatility forecasts for the SV, the SVX$^+$ and the SIV model where N denotes the forecasting horizon in terms of trading days. We then proceed with describing our two measures of realised volatility which are the squared daily returns and the intraday volatility measure together with our evaluation criteria.

6.4.1 Stochastic Volatility Model Forecasts

The one-step ahead volatility forecast for the SV model, as defined in equations (6.1), (6.2) and (6.3), is calculated as

$$\mathrm{E}(\sigma^2_{T+1|T}) = \exp(\ln \hat{\sigma}^{*2} + h_{T+1|T} + 0.5p_{T+1|T}), \tag{6.7}$$

and the N-step ahead volatility forecast is defined as

$$\begin{aligned}
&\mathrm{E}(\sigma^2_{T+1,T+N|T}) = \\
&\sum_{j=1}^{N} \exp(\ln \hat{\sigma}^{*2} + h_{T+j|T} + 0.5p_{T+j|T}) = \\
&\hat{\sigma}^{*2} \exp(h_{T+1|T} + 0.5p_{T+1|T}) + \\
&\hat{\sigma}^{*2} \sum_{j=2}^{N} \exp\left[\hat{\phi}^{j-1} h_{T+1|T} + 0.5\left(\hat{\phi}^{2(j-1)} p_{T+1|T} + \sum_{i=0}^{N-2} \hat{\phi}^{2i} \hat{\sigma}^2_{\eta}\right)\right],
\end{aligned} \tag{6.8}$$

where $\hat{\sigma}^{*2}$, $\hat{\phi}$ and $\hat{\sigma}^2_{\eta}$ are the maximum likelihood estimates of σ^{*2}, ϕ and σ^2_{η}, respectively. The estimator of h_{T+1} using all observations available at time T is denoted by $h_{T+1|T}$ with variance $p_{T+1|T}$ and both are computed by simulation methods which are discussed in Appendix B.

The multi-step forecasts are defined as a summation of the N individual forecasts conditional on the information available at time T. As N increases, the individual forecast $\mathrm{E}(\sigma^2_{T+N|T})$ will converge to a constant which we call the individual long-term volatility forecast and which is identical to the unconditional variance as given by

$$\hat{\sigma}^{*2} \exp\left(0.5 \frac{\hat{\sigma}^2_\eta}{1-\hat{\phi}^2}\right).$$

It is evident from equation (6.8) that the rate at which the individual forecasts move towards this value is determined by the size of $\hat{\phi}$; the smaller the volatility persistence estimate, the faster the individual forecasts converge to the individual long-term volatility forecast value. In empirical applications for daily stock returns the volatility persistence estimates are invariably found to be close to unity, which means that for shorter forecasting horizons individual forecasts are almost solely determined by the size of the short-term volatility, denoted by $\hat{\sigma}^{*2} \exp(h_{T+1|T} + 0.5 p_{T+1|T})$.

6.4.2 SVX$^+$ and SIV Model Forecasts

The one-step ahead forecasts for the SVX$^+$ model, as defined in equations (6.1), (6.2) and (6.5) are obtained in a similar way as for the SV model in equation (6.7) and using the same methods. This also applies to the SIV model with the h_t process defined in equation (6.6). For both the SVX$^+$ and SIV models we require x_{T+j-1} with $j = 1, \ldots, N$ for the N-period ahead volatility forecast in order to calculate the values for $h_{T+j|T}$ and $p_{T+j|T}$. However, for $N \geq 2$ this value is not known at time T. For $N \geq 2$ we therefore define the N-step ahead volatility forecasts of the SVX$^+$ and the SIV model as an N multiple of the one-step ahead volatility forecast, so we calculate these as

$$\mathrm{E}(\sigma^2_{T+1,T+N|T}) = N\hat{\sigma}^{*2} \exp(h_{T+1|T} + 0.5 p_{T+1|T}). \tag{6.9}$$

6.4.3 Measuring Predictive Forecasting Ability

To evaluate the accuracy of variance forecasts they have to be compared with realised volatility which cannot be observed. It is common practice in the literature to define the actual or realised variance as the squared observed returns, which for one-day ahead volatility is equal to

$$R^2_{T+1} = \sigma^2_{T+1} \varepsilon^2_{T+1}. \tag{6.10}$$

However, the squared error term ε_{T+1}^2 will vary widely which implies that only a relatively small part is attributable to σ_{T+1}^2.

An alternative approach which addresses this problem has been suggested ; see, for example, Andersen and Bollerslev (1998), Andersen, Bollerslev, Diebold and Labys (2001), Barndorff-Nielsen and Shephard (2001) and Andersen, Bollerslev, Diebold and Ebens (2001). It was first pointed out by Andersen and Bollerslev (1998) that more accurate ex-post volatility measures could be obtained with the sum of squared intraday returns where theoretically the estimates become free of measurement noise for infinitesimally small sampling frequency intervals. Following the above studies we define intraday squared returns as the overnight return plus the sum of the squared f-minute returns between 9.30 a.m. and 4.00 p.m. EST during the relevant trading day. As a full trading day has 391 minute-by-minute observations the daily sample consists of one overnight return and 390 divided by f intraday returns, so at sampling f the daily sample size is equal to $N_f = 1 + \frac{390}{f}$ where f is chosen so that N_f is an integer. The intraday volatility for day $T+1$ at sampling frequency f is therefore defined as

$$\begin{aligned} \tilde{\sigma}_{T+1,f}^2 &= [100(\ln P_{T+1,f,1)} - \ln P_{T,f,N_f)})]^2 + \\ & \sum_{m=2}^{N_f} [100(\ln P_{T+1,f,m} - \ln P_{T+1,f,m-1})]^2 \\ &= \sum_{m=1}^{N_f} R_{T+1,f,m}^2, \end{aligned} \tag{6.11}$$

where $P_{T+1,f,1}$ denotes the 9.30 a.m. price of the Standard & Poor's 100 Index on day $T+1$. The closing prices at day T and $T+1$ are given by P_{T,f,N_f} and P_{T+1,f,N_f}, respectively.

The multiple-day values of the daily squared returns and the intraday squared returns are obtained by summing the realised volatility measures of equations (6.10) and (6.11) over the relevant forecasting interval, so

$$R_{T+1,T+N}^2 = \sum_{i=1}^{N} R_{T+i}^2 \tag{6.12}$$

with

$$R_{T+i}^2 = [100(\ln P_{T+i} - \ln P_{T+i-1})]^2,$$

where P_{T+i} denotes the closing price of the Standard & Poor's Index at time $T+i$,
and

$$\tilde{\sigma}^2_{T+1,T+N,f} = \sum_{i=1}^{N} \tilde{\sigma}^2_{T+i,f}, \tag{6.13}$$

with

$$\begin{aligned} \tilde{\sigma}^2_{T+i,f} &= [100(\ln P_{T+i,f,1)} - \ln P_{T+i-1,f,N_f)})]^2 + \\ & \sum_{m=2}^{N_f} [100(\ln P_{T+i,f,m} - \ln P_{T+i,f,m-1})]^2. \end{aligned} \tag{6.14}$$

In order to assess the predictive accuracy of the volatility forecasts we compare the goodness-of-fit statistics, which are calculated from the regressions

$$R^2_{T+1,T+N} = a + b\,\mathrm{E}(\sigma^2_{T+1,T+N|T}) + \epsilon, \tag{6.15}$$

and

$$\tilde{\sigma}^2_{T+1,T+N,f} = a + b\,\mathrm{E}(\sigma^2_{T+1,T+N|T}) + \epsilon, \tag{6.16}$$

for the squared and the intradaily squared returns, respectively. If the relevant volatility forecast is unbiased, we have $a = 0$ and $b = 1$. This hypothesis can be jointly tested using the χ^2_2 Wald statistic .

In addition to the regression-based evaluation method we also report on a number of error statistics, which are the mean squared error (MSE), the median squared error (MedSE) and the mean absolute error (MAE) as these criteria are also frequently applied in the volatility forecasting literature.

6.4.4 Intraday Volatility

The high frequency data used for our out-of-sample intraday volatility measure covers the period 6 January 1997 to 29 June 2001. Initially we select a sampling frequency of five minutes, *i.e.* $f = 5$ where the latest stock index price available before the five-minute mark has been used for the calculation of the returns as defined in equation (6.11). Non-synchronised trading in the stock index can however induce serious positive autocorrelation in the stock index returns and therefore lead to underestimation of the actual volatility; the reverse then holds when the high frequency returns are negatively correlated which can be the result of bid-ask bounce in the stock index[14]. We find that our high frequency return series with $f = 5$ is autocorrelated at lag 1 with $\hat{\rho_1} = 0.07$ which is highly significant. The square root of the average annualised intraday volatility is equal to 19.4% which is much smaller than the annualised standard deviation of 21.4% calculated from the daily returns over the same period[15]. This suggests that actual volatility is indeed underestimated by the intraday volatility measure. Following French, Schwert and Stambaugh (1987) we could consider calculating intraday volatility as the sum of the squared 5-minute returns plus twice the sum of the cross-products of the adjacent 5-minute returns, *i.e.* as

$$\begin{aligned}\sum_{m=1}^{N_f} R^2_{T+1,f,m} \quad &+ \quad 2\sum_{m=1}^{N_f-1} R_{T+1,f,m}R_{T+1,f,m+1} = \\ \tilde{\sigma}^2_{T+1,f} \quad &+ \quad 2\sum_{m=1}^{N_f-1} R_{T+1,f,m}R_{T+1,f,m+1}. \qquad (6.17)\end{aligned}$$

For our sample the average daily value of the second term in the equation is equal to 0.28 which increases the intraday volatility measure to an annual average of 21.1%. A more acceptable method however is to decrease the sampling frequency and select the lowest value for f at which the summed cross-product term starts to approach zero and the autocorrelation has virtually disappeared from the sample. We calculated these values for $f = 5, 10, 15, 30, 65, 130, 195$ and 390 and present the results in figure 6.2. The first-order sample autocorrelation coefficients are lowest for the 10, 15 and 30-minute series but as the cross-product term increases for $f = 30$ we are left with two sampling frequencies. Although the 15-minute return series has a slightly lower value for the second term in equation (6.17) it only consists of two-thirds of the number of observations in the 10-minute series. Taking this into consideration we conclude that the 10-minute frequency is the most optimal for our purposes[16]. For the calculation of our realised volatility measure in equation (6.11) we therefore set f equal to 10. At this frequency $\hat{\rho_1} = 0.01$, the autocovari-

[14] See Campbell, Lo and MacKinlay (1997, Chapter 3).

[15] For the VIX data set the equivalent volatility is even higher at 26.0%.

[16] On the basis of the same criteria Oomen (2001) selected a sampling frequency for the FTSE 100 stock index of 25 minutes. We also considered a sample that excluded the overnight returns which had little effect on our results.

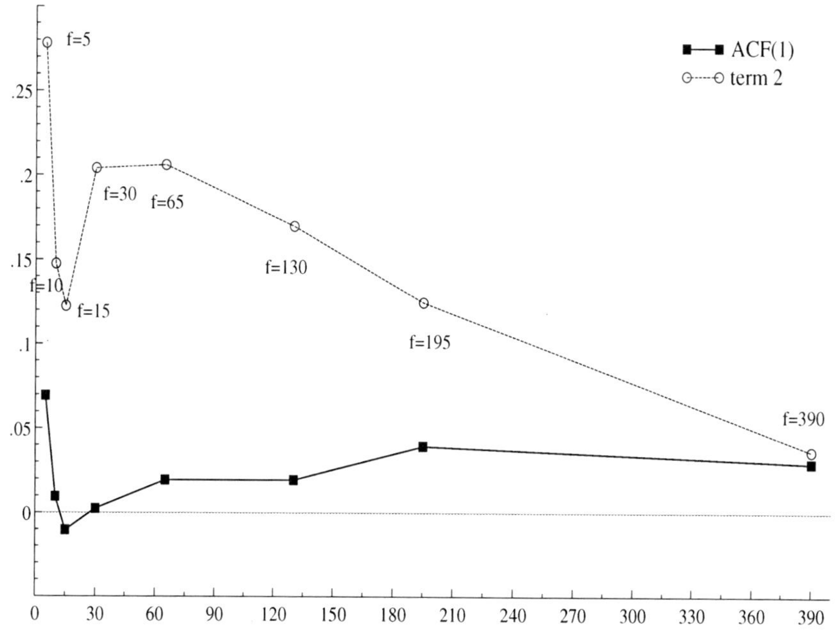

Figure 6.2. Sample first-order autocorrelation coefficients and summed cross-products multiplied by 2 as defined in equation (6.17) for sampling frequencies $f = 5, 10, 15, 30, 65, 130, 195$ and 390.

ance bias factor in equation (6.17) is reduced to 0.15 and the average annual volatility value has increased to 20.2% .

6.5 Out-of-Sample Results

In this section we report on the out-of-sample forecasting results of the SV, SVX$^+$ and SIV models over the evaluation period 6 January 1997 to 29 June 2001. Prior to the presentation and discussion of these results in section 6.5.2 we examine the relationships between the various parameters in the SV model.

6.5.1 The Parameters Estimates of the SV Model

For the one-day ahead SV volatility forecasts we had to estimate the SV model 1128 times which resulted in an equal number of estimates for all three parameters in the model. For illustrationary purposes in this subsection we expanded this set by starting the "fitting sample" at the beginning of 1986, hence we also calculated parameter estimates based on samples which included the 1987 stock market crash. In total we obtained 1632 estimates for each of the ϕ, σ_η^2 and σ^{*2} parameters as the forecasting sample rolled forward by one observation at the time; this now allows us to examine the dynamics of the SV model. In figure 6.3 we plot the estimates of the persistence parameter ϕ, the error variance of the volatility process σ_η^2, and the scaling parameter σ^{*2}. These estimates are based on the previous 9 years of data and the sample variance of each of these 1632 data series is plotted in the bottom-right-hand corner of the same figure.

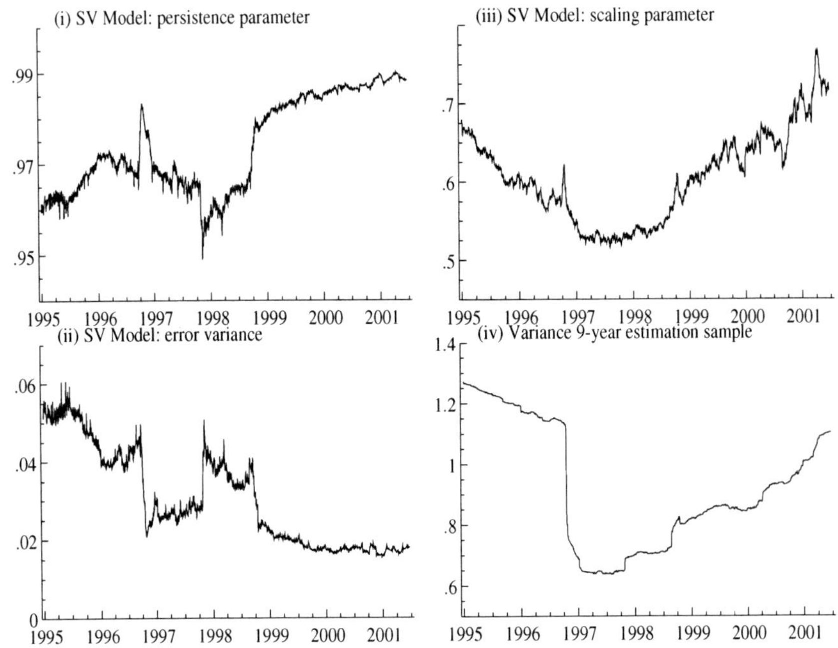

Figure 6.3. Parameter estimates (i) ϕ, (ii) σ_η^2 and (iii) σ^{*2} of the SV model and (iv) the variance of the estimation sample based on the previous 9 years of data.

The graph for the $\hat{\phi}$ parameter shows that volatility is highly persistent for all samples and when we compare this time series with that of $\hat{\sigma}_\eta^2$ we observe that the two series move in opposite directions during the entire period. The negative relationship between these two parameter estimates indicates that large (unexpected) shocks to the volatility process have a downward effect on the ϕ estimate and that volatility persistence is higher when these shocks are more moderate in size, *i.e.*

when values for $\hat{\sigma}_\eta^2$ are smaller. The variance of the estimation samples which is depicted in the bottom-right-hand corner of figure 6.3 shows a sharp decrease after approximately two years when the observations relating to the 1987 stock market crash drop out of the estimation sample. This decrease in volatility is also reflected in $\hat{\sigma}_\eta^2$ and in the estimated values for the scaling parameter σ^{*2} which displays a positive correlation with the variance series of the estimation samples over the full period. The graph for $\hat{\sigma}^{*2}$ also clearly shows a slowly but steadily increasing estimate for the scaling parameter during the second half of the sample. This suggests that in more recent years returns on the Standard & Poor's 100 index have become increasingly volatile which we know cannot be attributed to a few outliers (see graph 6.3-iv which increases gradually rather than abruptly). The relationship between the scaling parameter and the other two SV parameters then differs across the sample which is to be expected as they measure different aspects of the volatility process: the estimate for σ^{*2} reflects the level of volatility, $\hat{\phi}$ measures the degree of volatility persistence and the value for $\hat{\sigma}_\eta^2$ indicates the amount of variation in the volatility process. This means that during times of persistent high volatility we will observe high values for $\hat{\sigma}^{*2}$ and $\hat{\phi}$ but low values for $\hat{\sigma}_\eta^2$ as there is relatively little movement in the volatility process itself. However, the estimated value for the scaling parameter will still be large when a high level of volatility is due to a few outliers, but the variation in the volatility process as measured by $\hat{\sigma}_\eta^2$ is going to be higher and will be accompanied by a smaller value for the volatility persistence estimate .

6.5.2 Empirical Out-of-Sample Forecasting Results

The empirical out-of-sample forecasting results are based on the methodology described in section 6.4 and they are presented in tables 6.3 and 6.4.

In table 6.3 we evaluate the volatility forecasts obtained with the SV, SVX^+ and SIV models against the daily squared returns as defined in equation (6.12). The goodness-of-fit R^2 statistics indicate that the SV models which use the information contained in the VIX index consistently provide more accurate volatility forecasts than the standard SV model which only utilises the historical Standard & Poor's 100 returns therefore confirming the most recent findings in the GARCH literature, see for example Blair *et al.* (2001). The SVX^+ model is then marginally outperformed by the SIV model which has the highest R^2 values for all forecasting horizons considered. Although the Wald statistic for testing

Table 6.3. Out-of-sample forecasting results evaluated against daily squared returns for the (i) SV model, (ii) SVX$^+$ model and the (iii) SIV model based on the 1988–2001 sample and for the evaluation period 6 January 1997 to 29 June 2001

Forecasting Model		Forecasting Horizon N					
		1	2	5	10	15	20
SV Model	R^2	0.0372	0.0429	0.0664	0.0802	0.0847	0.0868
	Wald	17.962	15.408	14.687	18.257	24.781	32.512
	MSE	17.605	43.900	133.72	335.12	587.22	889.40
	MedSE	1.0159	2.4498	7.6933	21.694	48.066	85.337
	MAE	1.7974	2.8248	5.2779	9.1508	13.181	17.225
SVX$^+$ Model	R^2	0.0752	0.0946	0.1347	0.1655	0.1644	0.1496
	Wald	19.558	19.228	20.951	26.809	34.819	43.011
	MSE	17.514	43.237	130.53	321.58	562.85	861.27
	MedSE	0.7176	1.7987	5.6459	15.471	38.568	67.440
	MAE	1.6615	2.6378	4.9566	8.6202	12.534	16.557
SIV Model	R^2	0.0858	0.1046	0.1407	0.1702	0.1672	0.1543
	Wald	21.879	20.434	21.603	27.206	34.494	43.084
	MSE	17.420	42.988	129.96	320.35	561.54	857.69
	MedSE	0.7213	1.8185	5.4396	15.347	37.143	61.993
	MAE	1.6615	2.6338	4.9422	8.5822	12.483	16.477

Parameter estimates and goodness-of-fit R^2 statistics for the OLS regressions as defined in equation (6.15). Wald denotes the asymptotic χ_q^2 distributed Wald statistic based on standard errors using Newey-West heteroskedasticity and autocorrelation consistent covariance estimates testing for the joint null hypothesis $a = 0$ and $b = 1$; q is the number of restrictions under the null hypothesis and the critical value at the 1% significance level is 9.21. The highest values for R^2 and the lowest Wald and error statistic values are underlined.

the joint null hypothesis of $a = 0$ and $b = 1$ in equation (6.16) shows that the hypothesis of unbiased volatility forecasts has to be rejected for all three models and for all values of N, we find that on the basis of this "unbiasedness" criterion the SIV model is the *worst* performing model

Table 6.4. Out-of-sample forecasting results evaluated against the intraday volatility measure with $f = 10$ for the (i) SV model, (ii) SVX^+ model and the (iii) SIV model based on the 1988–2001 sample and for the evaluation period 6 January 1997 to 29 June 2001

Forecasting Model		Forecasting Horizon N: 1	2	5	10	15	20
SV Model	R^2	0.1824	0.2069	0.2250	0.2107	0.1891	0.1745
	Wald	14.061	19.842	24.304	30.327	37.074	43.163
	MSE	2.9159	8.3404	35.423	116.49	240.91	403.42
	MedSE	0.2521	0.8332	4.1678	15.526	31.741	55.897
	MAE	0.8829	1.5578	3.5162	6.6281	9.8185	13.174
SVX^+ Model	R^2	0.2550	0.2941	0.3182	0.2857	0.2523	0.2199
	Wald	36.616	35.686	34.264	37.505	43.258	48.327
	MSE	2.9336	8.2465	34.045	111.03	228.86	388.01
	MedSE	0.1474	0.4867	2.6420	10.551	26.198	45.399
	MAE	0.7944	1.4140	3.2448	6.2964	9.4296	12.565
SIV Model	R^2	0.2840	0.3167	0.3356	0.2981	0.2614	0.2284
	Wald	39.150	38.033	35.547	37.676	43.251	48.607
	MSE	2.8721	8.0919	33.465	109.56	226.58	384.34
	MedSE	0.1394	0.4679	2.5292	10.562	23.370	42.951
	MAE	0.7841	1.3974	3.2061	6.2231	9.3345	12.473

Parameter estimates and goodness-of-fit R^2 statistics for the OLS regressions as defined in equation (6.16). Wald denotes the asymptotic χ^2_q distributed Wald statistic based on standard errors using Newey-West heteroskedasticity and autocorrelation consistent covariance estimates testing for the joint null hypothesis $a = 0$ and $b = 1$; q is the number of restrictions under the null hypothesis and the critical value at the 1% significance level is 9.21. The highest values for R^2 and the lowest Wald and error statistic values are underlined.

and that the volatility forecasts of the SV model are the least biased[17].

[17] Poon and Granger (2001a) discuss the fact that as volatility is in practice estimated with a measurement error $E(\hat{b}) < 1$ and $E(\hat{a}) > 0$.

The reported error statistics confirm the earlier regression results which suggest that the SIV model is the preferred model for forecasting purposes, although the SVX^+ model appears to perform equally well. This last observation is not surprising given that $\hat{\phi}$ in the SVX^+ model is close to zero which implies that the forecasts of both models should be highly correlated[18]. On the basis of the results in table 6.3 we therefore conclude that the SV models which include implied volatility information appear to provide more accurate volatility forecasts than those that do not. However inclusion of the implied volatility measure also results in forecasts for which the hypothesis of unbiasedness has to be rejected at higher confidence levels. This might be due to the fact that only a relatively small part of the variation in the squared returns is attributable to the actual volatility process.

In table 6.4 we present the forecasting evaluation results based on realised returns defined in equation (6.13) with $f = 10$. In line with the earlier findings in the high frequency return literature, see *e.g.* Andersen and Bollerslev (1998), a considerable increase in the values for the R^2 statistic is observed for all volatility models and at all forecasting horizons when realised volatility is defined in terms of cumulative ten-minute squared returns. Again the volatility forecasts of the SIV model produces the highest values for the goodness-of-fit statistic and the lowest error statistics. Nevertheless the Wald statistics still indicate that the null hypothesis of unbiasedness has to be rejected with regard to all models and for all values of N. This problem is then most pronounced for the SVX^+ and the SIV model. The lowest Wald statistic values are those of the SV model where volatility forecasts are solely based on the information contained in the historical daily returns of the Standard & Poor's 100 index. The level of unbiasedness appears to increase as the forecasting horizon lengthens.

In order to investigate the volatility forecasts further we plot in graph 6.4 the one-period ahead volatility forecasts of the three models together with the daily squared returns and the intraday volatility measure with $f = 10$ over the full four-and-a-half year evaluation period. With the forecasting series of the three SV class models positioned on the right-hand-side and the realised volatility measures on the left-hand-side of the graph we observe that all the models produce out-of-sample one-day ahead volatility forecasts that are low during relatively quiet periods

[18] For all forecasting horizons volatility forecasts of the SVX^+ and SIV model are near perfectly correlated with sample correlation coefficients of 0.99.

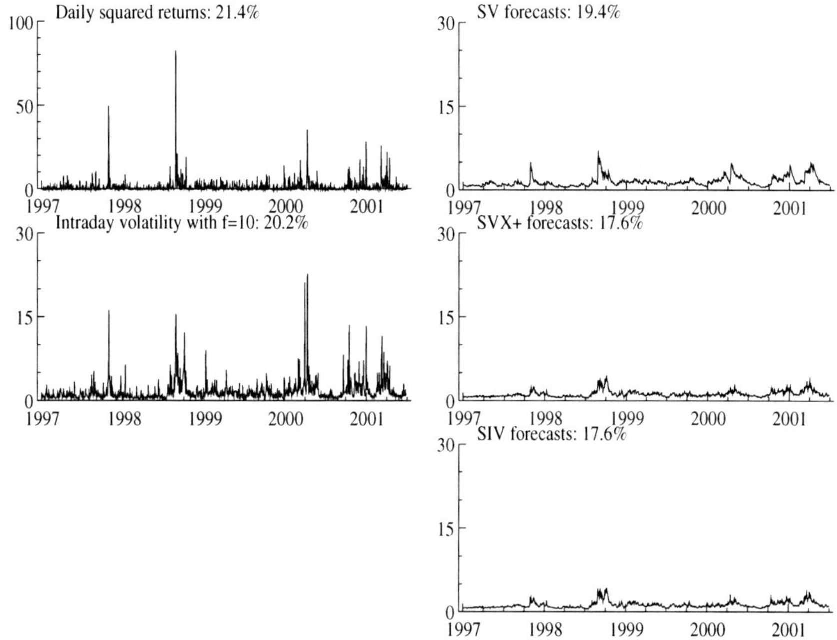

Figure 6.4. Daily squared returns, intraday volatility based on 10-minute squared returns and the VIX implied volatility together with the one-day ahead volatility forecasts of the SV, SVX^+ and SIV model for the Standard & Poor's 100 index over the period 06/01/97 to 29/06/01 based on a 9-year rolling window sample and with average annual volatilities given in percentages.

and high during the more volatile periods[19]. We calculated the average annual volatilities of the volatility series and found that all three SV models underestimated realised volatility on average[20]. We believe that the average underestimation by the SV models is closely related to the fact that the 9-year period preceding the 1997-2001 evaluation period is less volatile than the period over which we conduct the out-of-sample

[19]The standard SV model forecasts and the forecasts obtained with the SVX^+ and SIV model have sample correlation coefficients of 0.74 and 0.76, respectively.

[20]Had we used the VIX index directly the realised volatility process would have been overestimated on average, see footnote 15. Also see Fleming (1998) who reports that the Standard & Poor's 100 implied volatilities calculated separately from call and put options contain relevant information with regard to future volatility but are nevertheless upwardly biased.

forecasting evaluation (see graph 6.1)[21]. Although the VIX index itself produces volatility forecasts which on average are too high we demonstrated that when these implied volatility measures are embedded into the variance equation of the SV class of volatility model that they result in forecasts that are on average too low[22]. The volatility forecasts which appear to be least biased are then given by the SV model.

6.6 Summary and Conclusions

In this chapter we examine the predictive ability of Stochastic Volatility (SV) models and compare its forecasts with the volatility forecasts implied by option prices for daily returns on the Standard & Poor's 100 Stock Index. For this purpose we extend the SV model to a volatility model which allows for the inclusion of implied volatility as an exogenous variable in the variance equation. As the resulting SVX model includes an entire lag structure of the implied volatility measure we modify the SVX model to adjust for this with a persistence adjustment term. We refer to this model as the SVX^+ model. In addition we define the SIV model as a restricted volatility model for which the volatility process is determined by implied volatility and a stochastic error term. Our in-sample results indicate that historical returns are outperformed by implied volatility as we can never reject the null hypothesis of ϕ equal to zero. We find that the persistence adjustment term in the SVX^+ corrects for the overestimation of the implied volatility parameter and that its parameter estimates are very similar to those of the SIV model. Our results confirm earlier findings in the GARCH literature where recent research has indicated that implied volatility provides the most accurate volatility forecasts. The out-of-sample volatility forecasts are constructed for forecasting horizons ranging from 1 to 20 trading days and we approximate realised volatility as daily squared returns and cumulative intraday squared returns following research by, for example, Andersen and Bollerslev (1998). In order to correct for the underestimation of the volatility process due to significant positive first-order autocorrelation in the intraday 5-minute returns we choose a sampling interval of 10 minutes for the calculation of our intraday volatility mea-

[21] By comparison, the average annualised volatility of the 4 January 1988 to 3 January 1996 sample amounts to 13.2% and that of the evaluation sample to 19.9% when based on intraday volatility.

[22] It might be of interest for future research to investigate alternative methods for combining SV model volatility forecasts and option implied volatility. For example, in the GARCH literature Donaldson and Kamstra (2001) develop a model with trading volume as regime-switching variable. They find that implied volatility dominates during high volume periods and that GARCH is more important when trading volume is low.

sure. The relative forecasting accuracy of the various volatility models is evaluated using both regression-based evaluation methods and forecasting error statistics. The values of the R^2 statistics are consistently higher when we evaluate the volatility forecasts on the basis of intraday volatility. When we define realised volatility in terms of daily squared returns the SIV model outperforms both the SV and the SVX^+ model but the Wald test statistics indicate that the SV model produces volatility forecasts which are less biased than those with models which include implied volatility. When we use intraday squared returns for our forecasting evaluation we observe a similar pattern. On further examination we find that all three models produce volatility forecasts that are on average too low. We argue that the 9 year samples used for the parameter estimation are far less volatile than the evaluation sample and therefore the estimated scaling parameter $\hat{\sigma}^{*2}$ is too small for forecasting purposes in our study.

Chapter 7

STOCK INDEX VOLATILITY FORECASTING WITH HIGH FREQUENCY DATA

7.1 Introduction

Spurred by the initial research of Andersen and Bollerslev (1998) and Barndorff-Nielsen and Shephard (2001) high frequency intraday returns are increasingly considered for the purpose of approximating realised volatility. The notion that daily ex-post volatility is better approximated when based on cumulative squared intraday return data is supported by the theory that the measurement noise contained in daily squared returns prevents the observation of the actual volatility process but is reduced as the sampling frequency of the return series from which volatility is calculated is increased[1]. As such, it therefore theoretically justifies and extends the earlier work of French, Schwert and Stambaugh (1987), amongst others. Andersen and Bollerslev (1998) also showed that daily Generalised Autoregressive Conditional Heteroskedasticity (GARCH) volatility forecasts of exchange rates, when evaluated against intraday volatility measures, are far more accurate than had been previously assumed. These findings were subsequently confirmed with regard to stock index data by Blair, Poon and Taylor (2001) who examined the predictive accuracy of out-of-sample volatility forecasts based on GARCH models; we reached the same conclusion with regard to Stochastic Volatility (SV) models in Chapter 6.

As its value with regard to forecasting evaluation appears to have become generally recognised by now, research attention has shifted more towards the potential gains that might be obtained from using intraday

[1]See, for example, Andersen, Bollerslev, Diebold and Labys (2001a), Andersen, Bollerslev, Diebold and Labys (2001b), Barndorff-Nielsen and Shephard (2001) and Barndorff-Nielsen and Shephard (2002).

data as an information source for out-of-sample volatility forecasting. Andersen, Bollerslev and Lange (1999) examined whether the modelling of intraday returns could improve daily and perhaps even longer-run volatility forecasts. For forecasting horizons of 1 day, 1 week and 1 month they found that the most accurate GARCH out-of-sample volatility forecasts were always obtained with a 1-hour interval series[2]. Even though their theoretical study had indicated otherwise, empirical results deteriorated rather than improved as sampling intervals were shortened beyond this 1-hour mark. Rather than increasing the sampling frequency, Blair *et al.* (2001) suggested to incorporate the intraday volatility measure for stock index returns as an explanatory variable in the variance equation of a daily GARCH model. Like Andersen, Bollerslev and Lange (1999) they also observed a considerable improvement in the out-of-sample forecasting performance of the GARCH model. Martens (2001) then compared both GARCH-based methods for two exchange rates. He found that the most accurate intraday GARCH model, which proved to be the model with the highest sampling frequency, could not outperform the daily GARCH model extended with intraday volatility. These studies therefore indicate that intraday return series contain incremental information for longer-run volatility forecasts when used in combination with GARCH models, but thus far the issue has not been researched in the context of SV models. Here we examine whether extension of the SV model with intraday volatility information leads to similar improvements as observed for the GARCH model.

Alternatively, the intraday volatility process can be modelled directly which is reminiscent of the methods adopted for monthly volatility in a number of earlier studies such as those by French *et al.* (1987) and Poon and Taylor (1992). The forecasting performance of these volatility models have been studied by Ebens (1999) and Andersen *et al.* (2001b) who, in order to capture the long memory presumably present in the logarithms of intraday volatility series, prefer to use autoregressive fractionally integrated moving average (ARFIMA-RV) models with which they obtain more accurate forecasts than with daily GARCH models[3]. Following Barndorff-Nielsen and Shephard (2002) we define an alternative realised volatility model where volatility is modelled as a continuous time series process consisting of independent Ornstein-Uhlenbeck (OU)

[2]Due to strong periodic intraday patterns the GARCH models are not actually estimated at the higher intraday level, see *e.g.* Andersen and Bollerslev (1997). Instead the parameter estimates of the GARCH models at the higher frequencies are inferred from the temporal aggregation results of Drost and Nijman (1993).

[3]Fractionally integrated models are also advocated in the context of financial market data by, for example, Andersen, Bollerslev, Diebold and Labys (1999) and Andersen *et al.* (2001a).

processes. The resulting model for discrete time intervals is an unobserved components model which consists of independent ARMA components plus an error term and is estimated by casting it into state space form. We refer to this model as the Unobserved Components - Realised Volatility model, or, in short, the UC-RV model.

In this chapter we explore the forecasting performance of a number of models for the Standard & Poor's 100 stock index series over the period 6 January 1997 to 29 December 2000. We compare the UC-RV and the ARFIMA-RV model, which we collectively refer to as realised volatility models, with the so-called daily time-varying SV and GARCH volatility models which are also extended to include the intraday volatility measure. Thus far the emphasis in the high frequency volatility literature has been mainly on GARCH and ARFIMA-RV models. Our contribution is that we implement and empirically investigate a wide range of methods for the estimation of volatility to compare the out-of-sample forecasting performance of the various models.

The remainder of this chapter is organised as follows. In the next section we introduce the data together with intraday return based volatility measures. In section 7.3 we give details of the UC-RV and ARFIMA-RV models and in section 7.4 the daily time-varying volatility models are described. The forecasting methodology and the evaluation criteria are discussed in section 7.5 and in section 7.6 we present both the empirical in-sample and out-of-sample forecasting results. In section 7.7 we provide a summary and our conclusions.

7.2 Stock Return Data and Volatility

7.2.1 Data

The data for our empirical study consists of Standard & Poor's 100 stock index transaction prices during the period 6 January 1997 to 29 December 2000. From the original dataset, which includes prices recorded for every trade, we extract 5-minute interval data as this is the frequency used by Andersen and Bollerslev (1998) to construct their realised intraday volatility measure[4]. For the 5-minute price we take the last transaction price recorded before the relevant time mark and we calculate the 5-minute returns as the difference between successive log prices and express these in percentages, so

[4]Provided that the asset is sufficiently liquid, the 5-minute frequency is acknowledged as the highest frequency at which the effects of market microstructure biases, such as bid-ask bounces and discrete price observations, are not too distorting. Also see: Andersen *et al.* (2001a), Andersen *et al.* (1999), Ebens (1999) and Andersen and Bollerslev (1997).

$$R_{t,d} = 100(\ln P_{t,d} - \ln P_{t,d-1}), \tag{7.1}$$

where $R_{t,d}$ denotes the return for intraday period d on trading day t, with $d \geq 1$ and $t = 1, \ldots, T$. The New York Stock Exchange (NYSE) opens at 9.30 a.m. and closes at 4.00 p.m. EST. A full trading day therefore consists of 78 intraday returns and one overnight return. The overnight return is then defined in a similar way as the intraday returns, *i.e.*,

$$R_{t,N} = 100(\ln P_{t,0} - \ln P_{t-1,D}), \tag{7.2}$$

with $D = 78$, so $P_{t-1,D}$ is the 4.00 p.m. price on trading day $t-1$ and $P_{t,0}$ the 9.30 a.m. price on the following trading day t. For $P_{t,0}$ we always select the first available trading price after 9.29 a.m. Not all 1004 trading days in our sample consist of 79 observations and this is partly attributable to the fact that the NYSE closes early on certain days, such as on Christmas Eve[5]. Other important reasons are lapses in trading and in data reporting. For all these intervals without price quotes we insert zero return values except when the 9.30 a.m. price, which we require for the calculation of the overnight return, is not available[6]. In that case we assume that the first available price observation on day t, denoted by P_{t,d^*} with intraday period $d^* \geq 2$, was observed at 9.30 a.m. and assign a zero return value to R_{t,d^*}.

The daily return series R_t is defined as the first difference between the 4.00 p.m. closing prices on consecutive trading days, again expressed in percentage terms, so

$$R_t = 100(\ln P_t - \ln P_{t-1}), \tag{7.3}$$

where P_t and P_{t-1} could also be written as $P_{t,78}$ and $P_{t-1,78}$, respectively. In figure 7.1 we graph the daily return series R_t together with the squared return series R_t^2 over the full sample period and we report the summary statistics of both series in table 7.1. From the graphs we can discern several more volatile periods which occurred towards the end of 1997, during the third quarter of 1998 and at the beginning of 2000. Each

[5]Three trading days are missing from our data sample: 11 February 1998, 28 August 1998 and 14 January 1999.

[6]After accounting for early market closures, 559 price notations are "missing" of which 92 are 9.30 a.m. prices.

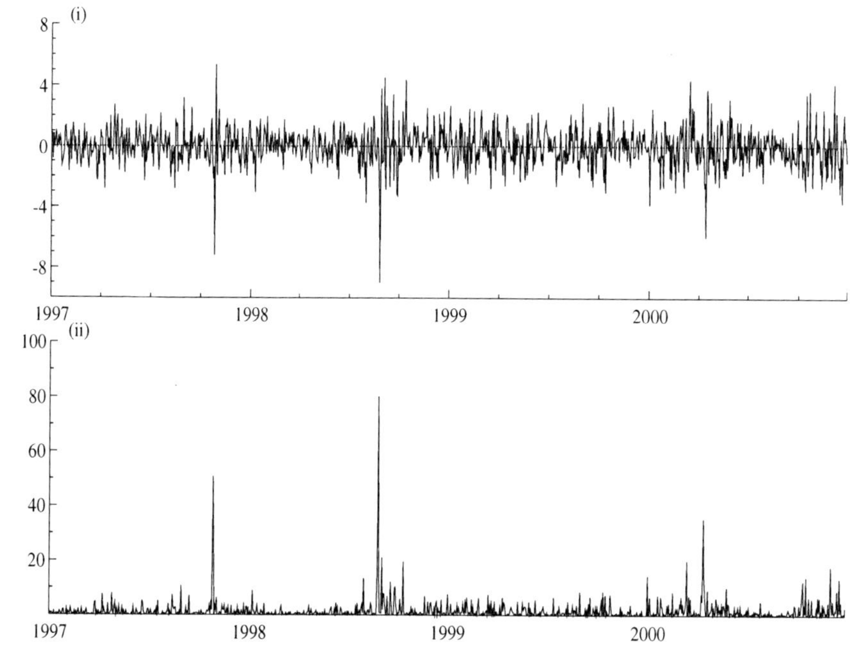

Figure 7.1. The daily (i) return series R_t and (ii) squared return series R_t^2 of the Standard & Poor's 100 stock index over the period 06/01/97 to 29/12/00

of the three largest shocks to the return process took place in one of these periods and was negative. This largely contributed to the negative skewness coefficient of -0.447 reported for the return series and the large positive skewness coefficient of 10.490 of the squared returns. We further observe that R_t exhibits excess kurtosis and that none of its first five autocorrelation coefficients is significantly different from zero at the 1% significance level. The Box-Ljung $Q(12)$ statistics indicate that returns are serially uncorrelated, whereas squared returns exhibit a high degree of serial correlation. The highly significant value for the first-order serial correlation coefficient $\hat{\rho}_1$ can be interpreted as an indication of volatility clustering. However, estimates for the subsequent three autocorrelation coefficients are not significant which might well be attributed to the fact that R_t^2 is a noisy estimator of the variance process of R_t.

7.2.2 Intraday Volatility

It has become generally acknowledged that squared daily returns provide a poor approximation of realised daily volatility. It was first pointed

Table 7.1. Summary Statistics R_t and R_t^2

Period	1997-2000	
No. of Obs. T	1004	
Series	R_t	R_t^2
Mean	0.063	1.673
Variance	1.670	16.576
Skewness	−0.447	10.490
Excess Kurtosis	4.024	164.548
Minimum	−8.947	0.000
Maximum	5.427	80.057
$\hat{\rho}_1$	−0.001	0.254
$\hat{\rho}_2$	−0.066	0.059
$\hat{\rho}_3$	−0.037	0.017
$\hat{\rho}_4$	0.010	0.042
$\hat{\rho}_5$	−0.040	0.122
$Q(12)$	21.780	93.224

$\hat{\rho}_\ell$ is the sample autocorrelation coefficient at lag ℓ with asymptotic standard error $1/\sqrt{T}$ and $Q(\ell)$ is the Box-Ljung portmanteau statistic based on ℓ squared autocorrelations.

out by Andersen and Bollerslev (1998) that more accurate estimates could be obtained with the sum of squared intraday returns[7]. More specifically, they defined realised volatility in the foreign exchange market as the sum of 288 5-minute squared returns. If we were to apply this method directly to the stock market, realised volatility would be defined as the sum of the squared overnight and the cumulative squared 5-minute intraday returns, so

$$\tilde{\sigma}_{t,1}^2 = R_{t,N}^2 + \sum_{d=1}^{D} R_{t,d}^2, \tag{7.4}$$

with $R_{t,d}$ and $R_{t,N}$ as defined in equations (7.1) and (7.2) and with $D = 78$. However, this ignores that the overnight return is a special case. Stock markets, unlike foreign exchange markets, are not opened

[7]Theoretically, the volatility estimates become free of measurement noise as the sampling frequency interval becomes infinitesimally small; see the references in note 1.

24 hours a day and the changes in the stock index price during the hours that the stock market is closed are relatively large compared to the 5-minute returns observed during trading hours. In order to account for the fact that overnight returns are presumably more volatile than intraday 5-minute returns and that a large value for $R_{t,N}$ will have a pronounced and distorting effect on the realised volatility estimate $\tilde{\sigma}^2_{t,1}$, we also use two alternative realised volatility measures which exclude the "noisy" overnight returns. The first of these is simply calculated as

$$\tilde{\sigma}^2_{t,2} = \sum_{d=1}^{D} R^2_{t,d}, \tag{7.5}$$

and therefore only measures the volatility during trading hours as opposed to *daily* volatility[8]. It was suggested by Martens (2002) to use a scaler in order to obtain a *daily* realised volatility measure based on intraday returns only. Furthermore, he found that in the absence of intranight returns the most accurate estimate of daily volatility on Standard & Poor's 500 index futures was obtained with this scaled sum of squared intraday returns, which he defined as

$$\begin{aligned} \tilde{\sigma}^2_{t,3} &= (1+c)\sum_{d=1}^{D} R^2_{t,d} \\ &= \frac{\sigma^2_{oc}+\sigma^2_{co}}{\sigma^2_{oc}}\sum_{d=1}^{D} R^2_{t,d}, \end{aligned} \tag{7.6}$$

where σ^2_{oc} and σ^2_{co} are the in-sample open-to-close and close-to-open variances, *i.e.* $\text{var}(\sum_{d=1}^{D} R_{t,d})$ and $\text{var}(R_{t,N})$, respectively, which implies $c \geq 0$.[9] For our full Standard & Poor's 100 stock index sample we find an open-to-close variance of 1.447 and a close-to-open variance of 0.092, resulting in a scaling value $(1+c)$ of 1.064 which is considerably lower than the value of 1.205 observed by Martens (2002) for his Standard & Poor's 500 stock index futures series.

In table 7.2 we provide summary statistics for $\tilde{\sigma}^2_{t,1}$ and $\tilde{\sigma}^2_{t,3}$, together with those of their logarithmic counterparts. We observe that the mean value of $\tilde{\sigma}^2_{t,1}$ is slightly higher than that of the scaled trading hours

[8]Andersen, Bollerslev, Diebold and Ebens (2001) and Andersen and Bollerslev (1997) use this definition of realised volatility in their stock market studies.

[9]Yet a different method is explored by Areal and Taylor (2002) who suggest assigning different weights to the intraday squared returns with weights depending on variance proportions which are calculated for each day of the week.

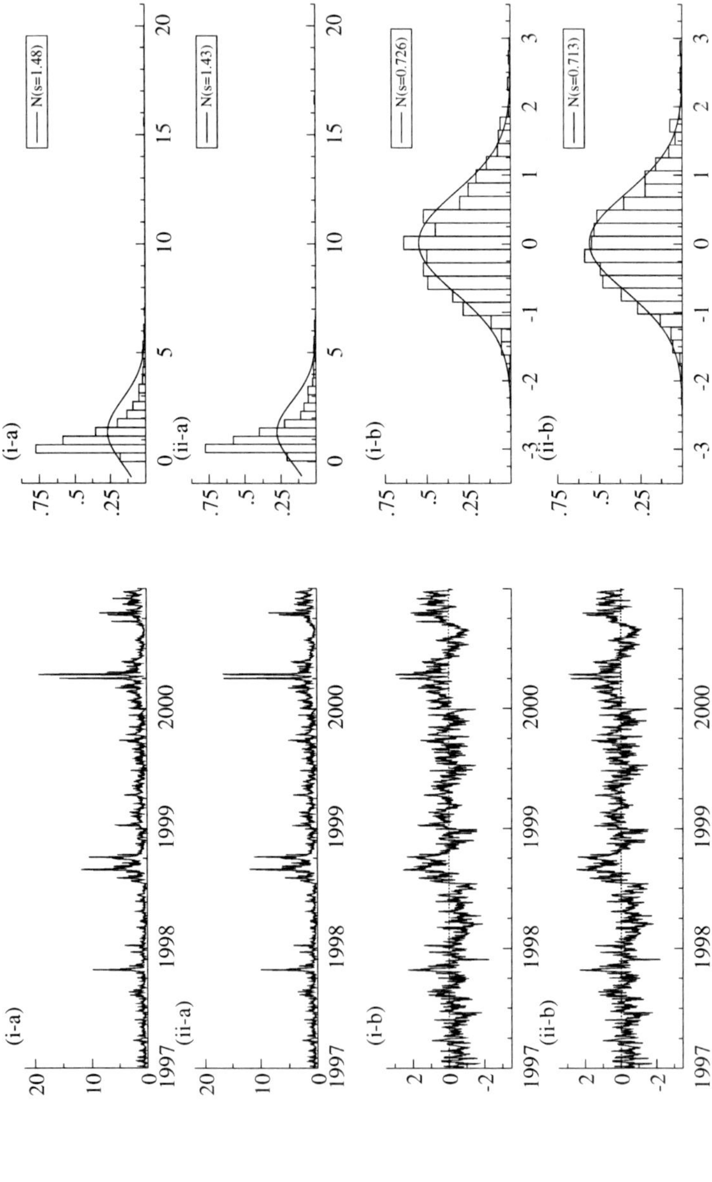

Figure 7.2. Time series and histograms with normal approximations for the Standard & Poor's 100 stock index realised volatility measures (i-a) $\tilde{\sigma}^2_{t,1}$ and (ii-a) $\tilde{\sigma}^2_{t,3}$ and their logarithmic counterparts (i-b) $\ln \tilde{\sigma}^2_{t,1}$ and (ii-b) $\ln \tilde{\sigma}^2_{t,3}$ over the period 06/01/97 to 29/12/00

Table 7.2. Summary statistics for the realised volatility measures

Period	1997-2000			
No. of obs. T	1004			
Series	$\tilde{\sigma}^2_{t,1}$	$\tilde{\sigma}^2_{t,3}$	$\ln\tilde{\sigma}^2_{t,1}$	$\ln\tilde{\sigma}^2_{t,3}$
Mean	1.372	1.359	0.017	0.020
Variance	2.207	2.053	0.527	0.508
Skewness	5.299	5.185	0.451	0.430
Excess Kurtosis	43.725	40.992	0.559	0.601
N	84678	74794	47.154	46.116
Minimum	0.106	0.109	−2.241	−2.215
Maximum	19.164	16.608	2.953	2.810
$\hat{\rho}_1$	0.588	0.572	0.622	0.609
$\hat{\rho}_2$	0.373	0.392	0.541	0.532
$\hat{\rho}_3$	0.318	0.323	0.502	0.484
$\hat{\rho}_4$	0.300	0.296	0.460	0.443
$\hat{\rho}_5$	0.296	0.261	0.427	0.400
$Q(12)$	1189.1	1155.2	2296.4	2120.8

N is the χ^2 normality test statistic with 2 degrees of freedom and a critical value of 9.21 at the 1% significance level. $\hat{\rho}_\ell$ is the sample autocorrelation coefficient at lag ℓ with asymptotic standard error $1/\sqrt{T}$ and $Q(\ell)$ is the Box-Ljung portmanteau statistic based on ℓ squared autocorrelations. The critical value at the 1% significance level for the Q(12) statistic is 26.22.

volatility measure $\tilde{\sigma}^2_{t,3}$ with values that translate into annualised standard deviations of 18.6% and 18.5%, respectively. The variance of $\tilde{\sigma}^2_{t,1}$ exceeds that of $\tilde{\sigma}^2_{t,3}$ which is explained by the fact that the latter does not include the noisy overnight return. Comparing the summary statistics of the intraday volatility measures with those of the daily squared return series in table 7.1 we find that R_t^2 has an average annualised standard deviation of 20.5% combined with a much higher degree of variation and larger skewness and excess kurtosis values. In addition, the reported autocorrelation coefficients for the intraday volatility series slowly decay as the lag length increases and they are always statistically significant, unlike those of the squared return series. As the intraday volatility series are clearly positively skewed and leptokurtic we also report on the logarithmic intraday volatility measures $\ln\tilde{\sigma}^2_{t,1}$ and $\ln\tilde{\sigma}^2_{t,3}$ as advocated in, for example, Andersen *et al.* (2001). The resulting four series are presented in figure 7.2 together with their histograms. The series $\tilde{\sigma}^2_{t,1}$ and $\tilde{\sigma}^2_{t,3}$ are highly correlated and their distributions are leptokurtic[10]. The distribution of the logarithmic intraday variances appears approximately Gaussian, mirroring earlier findings for the stock market by Ebens (1999), Andersen *et al.* (2001) and Areal and Taylor (2002). These conclusions are further supported by the skewness and excess kurtosis coefficients which have standard errors equal to $\sqrt{6/T} = 0.077$ and $\sqrt{24/T} = 0.155$, respectively, and are therefore close to normal but still have values which are different from zero at very high significance levels[11]. The sample autocorrelation coefficients indicate a highly persistent volatility process and the values for the $Q(12)$ statistic are so high that the hypothesis of zero autocorrelation is convincingly rejected[12].

7.3 Realised Volatility Models

In this section we discuss the two realised volatility models which are the unobserved components (UC-RV) and the ARFIMA (ARFIMA-RV) models where realised volatility is modelled directly as opposed to the volatility models in the next section which model volatility as the second moment of returns. Unlike the UC-RV model, the ARFIMA-RV model is defined in logarithmic terms.

7.3.1 Unobserved Components OU Type Stochastic Volatility Models

Realised volatility, as defined in subsection 7.2.2 and denoted by $\tilde{\sigma}^2_t$, can be used as an estimator of volatility σ^2_t; see, for example, Andersen and Bollerslev (1998). Barndorff-Nielsen and Shephard (2002) provide an excellent discussion of the properties of realised volatility. In particular, they investigate the statistical properties of the estimation error $\sigma^2_t - \tilde{\sigma}^2_t$ and they argue that a more accurate estimator can be obtained when a model is considered for σ^2_t. Actual volatility can be modelled as a continuous time series process consisting of independent Ornstein-Uhlenbeck (OU) processes, that is

[10] We find $corr(\tilde{\sigma}^2_{t,1}, \tilde{\sigma}^2_{t,3}) = 0.973$ and $corr(\ln \tilde{\sigma}^2_{t,1}, \ln \tilde{\sigma}^2_{t,3}) = 0.985$.

[11] The t-statistics for the skewness coefficients are 5.83 and 5.56 and those of the excess kurtosis coefficients are 3.62 and 3.89 for $\ln \tilde{\sigma}^2_{t,1}$ and $\ln \tilde{\sigma}^2_{t,3}$, respectively.

[12] Also see Ebens (1999) and Andersen *et al.* (2001) for similar autocorrelation coefficients and comparable Q statistic values.

$$\sigma^2(t) = \sum_{j=1}^{J} \tau^j(t), \qquad \mathrm{d}\tau^j(t) = -\lambda_j \tau^j(t)\mathrm{d}t + \mathrm{d}z_j(\lambda_j t),$$

where $\sigma^2(t)$ is the continuous-time process for actual volatility, $z_j(t)$ are independent Lévy processes with non-negative increments and λ_j are unknown parameters, for $j = 1, \ldots, J$ and J is typically a small number in practice (say, 3 or 4). Developments of this approach, with special attention to statistical and probabilistic aspects, are reported by Barndorff-Nielsen and Shephard (2001).

The formulation of the continuous time model for $\sigma^2(t)$ implies that actual volatility in discrete time intervals of length Δ can be modelled as

$$\sigma_t^2 = \sum_{j=1}^{J} \tau_t^j, \qquad t = 1, \ldots, T,$$

where each τ_t^j represents an ARMA(1,1) model with the autoregressive and moving average coefficients determined by $\exp(\lambda_j \Delta)$; see Barndorff-Nielsen and Shephard (2002) for further details. A model for realised volatility is then simply given by

$$\tilde{\sigma}_t^2 = \sigma_t^2 + u_t, \qquad t = 1, \ldots, T,$$

where u_t is white noise, with mean zero and variance depending on Δ, and is uncorrelated with σ_j^2 for $j = 1, \ldots, T$. The resulting model is an unobserved components model which consists of J independent ARMA components plus an error component. Linear optimal estimators of this model can be obtained by casting the model into state space form. The Kalman filter can be applied to construct the Gaussian likelihood function. Quasi-maximum likelihood estimates of λ_j are obtained by numerically maximising the Gaussian likelihood with respect $\lambda_1, \ldots, \lambda_J$.

In this chapter we adopt this model for realised volatility $\tilde{\sigma}_t^2$ with $J = 1$. However, we estimate the parameters of the ARMA process freely without taking allowance for its dependence on Δ. The resulting model will be referred to as the Unobserved Components - Realised Volatility, or UC-RV, model and is given by

$$\begin{aligned} \tilde{\sigma}_t^2 &= \sigma_t^2 + u_t, \\ \sigma_t^2 &= \mu + \phi(\sigma_{t-1}^2 - \mu) + \theta\eta_{t-1} + \eta_t. \end{aligned}$$

Note that we take the variance of u_t as implied by Δ but we do estimate the variance of η_t, that is σ_η^2. This leads to a model with four unknown parameters: μ, ϕ, θ and σ_η. The state space form of this model can be written as

$$\begin{aligned} \tilde{\sigma}_t^2 &= Z\alpha_t + u_t, \\ \alpha_{t+1} &= T\alpha_t + R\eta_t, \end{aligned} \tag{7.7}$$

where

$$Z = (1 \quad 0 \quad 0), \qquad T = \begin{bmatrix} \phi & 1 & 1 \\ 0 & 0 & 0 \\ 0 & 0 & 1 \end{bmatrix}, \qquad R' = (1 \quad \theta \quad 0),$$

and ϕ, θ and σ_u are the parameters of the ARMA(1,1) process. The first element of the 3×1 state vector α_t is $\tilde{\sigma}_t^2 - \mu$ and its last element is μ.

Estimation

When the ARMA(1,1) plus error model is casted in the state space model (7.7), the Kalman filter can be used to compute the Gaussian likelihood function and parameter estimates can be obtained by numerically optimising the resulting quasi-likelihood function. This methodology is standard for Gaussian linear models and is discussed extensively in, amongst others, Harvey (1989) and Durbin and Koopman (2002).

Forecasting

Forecasting for linear time series models in state space form is relatively straightforward. Firstly, the realised volatility series is artificially extended with missing values at the end. Secondly, the Kalman filter can deal with missing observations and is therefore applied to the new extended series. The estimates of the signal σ_t^2, corresponding to the missing values at $t = T+1, T+2, \ldots$, are taken as the forecasts of actual volatility σ_{T+j}^2, for $j = 1, 2, \ldots$. The mean squared errors of the forecasts are also provided by the Kalman filter. Details of this approach to forecasting are discussed in Durbin and Koopman (2002).

7.3.2 ARFIMA Models

In empirical work on realised volatility it is pointed out that the realised volatility series $\tilde{\sigma}_t^2$ can be regarded as being generated by a normal distribution after it is transformed by taking logarithms. The dynamic properties of log realised volatility exhibit features known as long memory, that is, the correlogram of such a series dies down slowly as the lag length increases. An appropriate model framework to deal with such specific dynamic properties is based on the ARFIMA model.

The ARFIMA$(1, d, 1)$ model with mean μ is given by

$$(1 - \phi L)(1 - L)^d (y_t - \mu) = (1 + \theta L)\varepsilon_t,$$

where L is the lag operator ($Ly_t = y_{t-1}$), coefficients d, ϕ and θ are fixed and unknown and ε_t is Gaussian white noise with mean zero and variance σ^2. The following restrictions on the parameters apply,

$$0 < d < 0.5, \quad |\phi| < 1, \quad |\theta| < 1, \quad \sigma^2 > 0.$$

In the context of volatility modelling, the ARFIMA model for the logs of realised volatility is empirically investigated by Ebens (1999), Andersen *et al.* (2001b) and Oomen (2001).

Estimation

The parameters of the ARFIMA model, including mean μ, can be estimated by the method of maximum likelihood; for details, see, for example, Sowell (1992). It is, for example, pointed out by Brodsky and Hurvich (1999) and Bos, Franses and Ooms (2002) that standard ARMA(1,1) models can also capture long-memory features and that, depending on the sample spectrum of the data, not all parameters of an ARFIMA(1,d,1) can be identified from the data. This usually applies to the case of realised volatility. In empirical studies one may fix the d parameter to a certain value and estimate the remaining parameters. However, we rather concentrate on the estimation of the crucial d parameter in an ARFIMA(1,d,0), ARFIMA(0,d,1) or ARFIMA(0,d,0) model. Although in prior analyses the three ARFIMA models produced rather similar results, we will consider the ARFIMA(1,d,0) model in our empirical study. The required computations are implemented using the Arfima package of Doornik and Ooms (2001) within the programming environment of Ox; see Doornik (1998).

Forecasting

Forecasting can be carried out by extrapolating the series in which the correlation structure implied by the estimated ARFIMA model is taken into account. Details of how these computations can be implemented elegantly for ARFIMA models are given by Doornik and Ooms (2001).

7.4 Daily Time-Varying Volatility Models

In this section we discuss the daily time-varying volatility models where volatility is explicitly modelled as the second moment of daily returns. This rather well-established range of volatility models includes the SV and GARCH classes of models. In addition to standard formulations we also consider extensions for both models with intraday volatility incorporated in the variance equation.

7.4.1 Daily SV Model

The daily Stochastic Volatility (SV) model can be defined as

$$\begin{aligned} y_t &= \sigma_t \varepsilon_t, & \varepsilon_t &\sim \text{NID}(0,1), \quad t = 1,\ldots,T, \\ \sigma_t^2 &= \sigma^{*2}\exp(h_t), & & \\ h_t &= \phi h_{t-1} + \sigma_\eta \eta_t, & \eta_t &\sim \text{NID}(0,1), \end{aligned} \tag{7.8}$$

where y_t denotes the return series of interest which is the daily series R_t as given in equation (7.3). The volatility process σ_t^2 is defined as the product of a scaling factor $\sigma^{*2} > 0$ and the exponential of the stochastic process h_t, which in turn is modelled as a first-order autoregressive process. The persistence parameter ϕ is restricted to be positive and smaller than one to ensure the stationarity of σ_t^2. We further assume that ε_t and η_t are mutually uncorrelated, both contemporaneously and at all lags. For reviews of the SV model we refer to Taylor (1994), Ghysels, Harvey and Renault (1996) and Shephard (1996).

Estimation

The parameters of the SV model are estimated by exact maximum likelihood methods using Monte Carlo importance sampling techniques. The likelihood function for the SV model can be constructed using simulation methods developed by Shephard and Pitt (1997) and Durbin and Koopman (1997). We start by considering the standard SV model of equation (7.8). The non-linear relation between log-volatility h_t and the observation equation of y_t does not allow the computation of the likelihood by linear methods such as the Kalman filter. For the SV model we can express the likelihood function as

$$L(\psi) = p(y|\psi) = \int p(y,\theta|\psi)d\theta = \int p(y|\theta,\psi)p(\theta|\psi)d\theta, \tag{7.9}$$

where

$$\psi = (\phi, \sigma_\eta, \sigma_\varepsilon)', \qquad \theta = (h_1, \dots, h_T)'.$$

An efficient way of evaluating such expressions is by using importance sampling; see Ripley (1987, Chapter 5). A simulation device is required to sample from an importance density $\tilde{p}(\theta|y,\psi)$ which we prefer to be as close as possible to the true density $p(\theta|y,\psi)$. An obvious choice for the importance density is the conditional Gaussian density since in this case it is relatively straightforward to sample from $\tilde{p}(\theta|y,\psi) = g(\theta|y,\psi)$ using simulation smoothers such as the ones developed by de Jong and Shephard (1995) and Durbin and Koopman (2002). Guidelines for the construction of an importance model and the likelihood function for the SV model using this approach are given in Appendix B.

Forecasting

For the case that forecasting horizon is one day, the daily volatility forecast for the SV model can be written as

$$\mathrm{E}(\sigma^2_{T+1|T}) = \hat{\sigma}^{*2} \exp(\hat{h}_{T+1|T} + 0.5 p_{T+1|T}), \tag{7.10}$$

where $\hat{\sigma}^{*2}$ is the maximum likelihood estimate of σ^{*2}, $\hat{h}_{T+1|T}$ is the estimator of h_{T+1} given all T observations and $p_{T+1|T}$ is its mean square error. When the forecasting horizon spans N days, we have

$$\mathrm{E}(\sigma^2_{T+1,T+N|T}) = \hat{\sigma}^{*2} \sum_{j=1}^{N} \exp(h_{T+j|T} + 0.5 p_{T+j|T}), \tag{7.11}$$

The estimator of h_{T+1} given all T observations, and its mean square error $p_{T+1|T}$ are computed with the simulation methods developed by Durbin and Koopman (2000) and applied to SV models in Chapter 6; for $j \geq 2$ the values for $\hat{h}_{T+j|T}$ and $p_{T+j|T}$ are given by

$$\begin{aligned}
\hat{h}_{T+j|T} &= \hat{\phi}^{j-1} \hat{h}_{T+1|T}, \\
p_{T+j|T} &= \hat{\phi}^{2(j-1)} p_{T+1|T} + \sum_{i=0}^{N-2} \hat{\phi}^{2i} \hat{\sigma}^2_\eta,
\end{aligned}$$

where $\hat{\phi}$ and $\hat{\sigma}^2_\eta$ are the maximum likelihood estimates of ϕ and σ^2_η, respectively. When these definitions of $\hat{h}_{T+j|T}$ and $p_{T+j|T}$ are considered in conjunction with equation (7.11) it becomes apparent that as N

increases $\mathrm{E}(\sigma^2_{T+N|T})$ will converge to a value which is identical to the unconditional variance given by

$$\hat{\sigma}^{*2} \exp\left(0.5 \frac{\hat{\sigma}^2_\eta}{1-\hat{\phi}^2}\right),$$

where the rate of convergence depends on the value for the volatility persistence parameter estimate $\hat{\phi}$ which tends to be close to unity for the type of daily financial time series we are studying here. Volatility forecasts for small values of N are therefore mainly determined by the values for $\hat{h}_{T+1|T}$ and $p_{T+1|T}$.

7.4.2 Daily SV Model with Intraday Volatility

The daily SV model with intraday volatility as explanatory variable in the variance equation is an extension of the SV model as defined in equation (7.8) where the stochastic process h_t is given by

$$h_t = \phi h_{t-1} + \gamma x_{t-1} + \sigma_\eta \eta_t \tag{7.12}$$

and where x_t denotes $\ln \tilde{\sigma}^2_t$ as defined in subsection 7.2.2. The value of ϕ we restrict to be smaller than one in absolute terms, so $-1 < \phi < 1$. Alternative formulations for the h_t process are possible, for example, one could consider the inclusion of a persistence adjustment term. However, we have chosen this one as it is most closely related to other models used in the intraday volatility literature[13]. The SV model with intraday volatility is identical to the SVX model introduced in Chapter 6 where we used implied volatility instead of intraday volatility as explanatory variable in the variance equation.

Estimation

The inclusion of realised volatility in the equation of h_t does not affect the non-linear relationship between observation y_t and the unobserved volatility component h_t. Therefore the estimation and forecasting methods for the SV model as described in subsection 7.4.1 can be applied straightforwardly. More details of maximum likelihood estimation using importance sampling techniques for the model in equation (7.12) are given in Appendix B.

[13]See *e.g.* Martens (2001) and Blair *et al.* (2001).

Forecasting

The daily forecast for the SV model with intraday volatility is obtained with similar methods as the daily forecast for the SV model defined in equation (7.10) but requires in addition the availability of x_T for the calculation of $\hat{h}_{T+1|T}$ and $p_{T+1|T}$. Because intraday volatility values are not known beyond time T we cannot calculate $\hat{h}_{T+j|T}$ and $p_{T+j|T}$ for $j \geq 2$ as we did in the case of the SV model. We therefore assume that these values do not change during subsequent trading days and define the equivalent of equation (7.11) as the one-day ahead volatility forecast multiplied by the length of the forecasting horizon N, so

$$\begin{aligned} \mathrm{E}(\sigma^2_{T+1,T+N|T}) &= N\hat{\sigma}^{*2}\exp(h_{T+1|T} + 0.5p_{T+1|T}) \\ &= N\,\mathrm{E}(\sigma^2_{T+1|T}). \end{aligned} \tag{7.13}$$

7.4.3 Daily GARCH(1,1) Model

The second class of daily time-varying volatility models we consider is the GARCH(1,1) model as given by

$$\begin{aligned} y_t &= \sigma_t \varepsilon_t \qquad \varepsilon_t \sim \mathrm{NID}(0,1), \qquad t = 1, \ldots, T, \\ \sigma_t^2 &= \omega + \alpha(\sigma_{t-1}\varepsilon_{t-1})^2 + \beta\sigma^2_{t-1}, \end{aligned} \tag{7.14}$$

with y_t as defined above in subsection 7.4.1 and parameter restrictions $\omega > 0$, $\alpha \geq 0$, $\beta \geq 0$ and $\alpha + \beta < 1$.

Estimation

Maximum likelihood methods for GARCH models are well established; the procedures developed by Bollerslev (1986) have been implemented in many standard econometric software packages such as PcGive and EViews[14]. Hence we do not give estimation details and refer to GARCH surveys, such as those by Bollerslev, Chou and Kroner (1992), Bera and Higgins (1993) and Bollerslev, Engle and Nelson (1994).

[14] More specifically, the parameters in the GARCH model were estimated using the G@RCH 1.1 package of Laurent and Peters (2002).

Forecasting

As all information for the one-day ahead volatility forecast is available at time T the daily GARCH(1,1) forecast can be directly calculated as

$$\mathrm{E}(\sigma^2_{T+1|T}) = \hat{\omega} + \hat{\alpha}\sigma^2_T\varepsilon^2_T + \hat{\beta}\sigma^2_T, \tag{7.15}$$

where $\hat{\omega}$, $\hat{\alpha}$ and $\hat{\beta}$ denote the maximum likelihood estimates of ω, α and β, respectively. The N-period ahead GARCH(1,1) forecast is then obtained by applying the law of iterated expectations and this forecast can be expressed as

$$\mathrm{E}(\sigma^2_{T+1,T+N|T}) = \sum_{j=1}^{N} \frac{\hat{\omega}}{1-\hat{\alpha}-\hat{\beta}} + (\hat{\alpha}+\hat{\beta})^{j-1}\left(\mathrm{E}(\sigma^2_{T+1|T}) - \frac{\hat{\omega}}{1-\hat{\alpha}-\hat{\beta}}\right). \tag{7.16}$$

From this equation we can deduce that $\mathrm{E}(\sigma^2_{T+N|T})$ converges to the unconditional variance value

$$\frac{\hat{\omega}}{1-\hat{\alpha}-\hat{\beta}}$$

as N increases and that the rate of convergence is governed by the sum of $\hat{\alpha}$ and $\hat{\beta}$, which measures the degree of volatility persistence.

7.4.4 Daily GARCH(1,1) Model with Intraday Volatility

The daily GARCH(1,1) model is extended to include intraday volatility by incorporating this information in the variance equation, so the volatility process σ^2_t in equation (7.14) can be rewritten as

$$\sigma^2_t = \omega + \alpha(\sigma_{t-1}\varepsilon_{t-1})^2 + \beta\sigma^2_{t-1} + \gamma x_{t-1}, \tag{7.17}$$

where x_t represents the realised volatility measure $\tilde{\sigma}^2_t$ as defined in subsection 7.2.2. We refer to this model as the GX model.

Estimation

Standard packages have options to include explanatory variables within the GARCH process as the estimation of the coefficients in equation (7.17) is relatively straightforward.

Forecasting

As for the standard GARCH model, all information for the calculation of the one-day ahead volatility forecast $\mathrm{E}(\sigma^2_{T+1|T})$ is available at time T as it is given by

$$\mathrm{E}(\sigma^2_{T+1|T}) = \hat{\omega} + \hat{\alpha}\sigma^2_T\varepsilon^2_T + \hat{\beta}\sigma^2_T + \hat{\gamma}x_T, \tag{7.18}$$

where $\hat{\gamma}$ is the maximum likelihood estimate of γ. Due to the fact that intraday volatility values are not known beyond time T we define the N-period ahead volatility forecast for the extended GARCH(1,1) model in a similar manner as for the SVX model, *i.e.*, as

$$\begin{aligned} \mathrm{E}(\sigma^2_{T+1,T+N|T}) &= N(\hat{\omega} + \hat{\alpha}\sigma^2_T\varepsilon^2_T + \hat{\beta}\sigma^2_T + \hat{\gamma}x_T) \\ &= N\,\mathrm{E}(\sigma^2_{T+1|T}). \end{aligned} \tag{7.19}$$

7.5 Forecasting Methodology and Evaluation Criteria

7.5.1 Forecasting Methodology

In the out-of-sample forecasting study below we initially estimate the parameters of each of the six volatility models described in sections 7.3 and 7.4 over the first 800 trading days of our sample, *i.e.*, over the period 6 January 1997 to 10 March 2000. The obtained parameter estimates are then inserted in the relevant forecasting formulas given in the same sections and volatility forecasts are calculated for horizons ranging from 1 day to 1 week, so for $N = 1, \ldots, 5$. The sample is then rolled forward by one trading day keeping the size of the sample constant at 800 observations and this procedure is repeated until we have obtained forecasts for the entire 13 March 2000 to 29 December 2000 out-of-sample period which consists of 204 trading days. As our volatility forecasts are overlapping, we have forecasting samples containing $205 - N$ volatility forecasts.

7.5.2 Evaluation Criteria

As a measure of realised volatility we use $\tilde{\sigma}^2_{t,3}$, which is the scaled intraday volatility measure defined in equation (7.6). We prefer this realised volatility measure to the highly correlated $\tilde{\sigma}^2_{t,1}$ of equation (7.4) as the latter includes noisy overnight returns. Realised volatility for forecasting horizons exceeding one trading day are calculated by summing the realised volatility measures of equation (7.6) over the relevant forecasting horizon N, so

$$\tilde{\sigma}^2_{(T+1,T+N),3} = \sum_{i=1}^{N} \tilde{\sigma}^2_{T+i,3}. \tag{7.20}$$

In order to assess the predictive abilities of the various volatility models we report on the goodness-of-fit coefficient R^2 as calculated from the OLS regression

$$\tilde{\sigma}^2_{(T+1,T+N),3} = a + b\,\mathrm{E}(\sigma^2_{T+1,T+N|T}) + \epsilon, \tag{7.21}$$

where $\mathrm{E}(\sigma^2_{T+1,T+N|T})$ denotes the N-period ahead volatility forecasts obtained with the volatility models defined in sections 7.3 and 7.4. If the volatility forecasts are unbiased, then $a = 0$ and $b = 1$. We test these hypotheses using standard regression methods with Newey-West adjustments to account for the special error covariance structure due to the rolling window construction of the forecasts. In addition to the regression-based method, which is by far the most popular post-sample evaluation procedure, we also report on two error statistics which are used by Andersen, Bollerslev and Lange (1999) and Martens (2002). These are the heteroskedasticity adjusted root mean squared error (HRMSE) and the mean absolute error (HMAE) which are calculated as

$$HRMSE = \sqrt{\frac{1}{205-N}\sum_{t=1}^{205-N}\left\{1 - \frac{\mathrm{E}(\sigma^2_{T+1,T+N|T})}{\tilde{\sigma}^2_{(T+1,T+N),3}}\right\}^2}, \tag{7.22}$$

and

$$HMAE = \frac{1}{205-N}\sum_{t=1}^{205-N}\left|1 - \frac{\mathrm{E}(\sigma^2_{T+1,T+N|T})}{\tilde{\sigma}^2_{(T+1,T+N),3}}\right|. \tag{7.23}$$

7.6 Empirical Results

7.6.1 In-Sample Results

In this subsection the results obtained with volatility models described in sections 7.3 and 7.4 are presented for the full in-sample period[15]. More specifically, we report on the estimation results of the following models:

UC-RV Model:
$$\tilde{\sigma}_{t,3}^2 = \sigma_t^2 + u_t,$$
$$\sigma_t^2 = \mu + \phi(\sigma_{t-1}^2 - \mu) + \theta\eta_{t-1} + \eta_t$$

ARFIMA-RV Model:
$$(1 - \phi L)(1 - L)^d \ln \tilde{\sigma}_{t,3}^2 = \mu + u_t$$

SV Model:
$$y_t = \sigma_t \varepsilon_t,$$
$$\ln \sigma_t^2 = (1 - \phi) \ln \sigma^{*2} + \phi \ln \sigma_{t-1}^2 + \sigma_\eta \eta_t$$

SVX Model:
$$y_t = \sigma_t \varepsilon_t,$$
$$\ln \sigma_t^2 = (1 - \phi) \ln \sigma^{*2} + \phi \ln \sigma_{t-1}^2 + \gamma_i \ln \tilde{\sigma}_{t-1,i}^2 + \sigma_\eta \eta_t$$

GARCH(1,1) Model:
$$y_t = \sigma_t \varepsilon_t,$$
$$\sigma_t^2 = \omega + \alpha(\sigma_{t-1}^2 \varepsilon_{t-1}^2) + \beta \sigma_{t-1}^2$$

GX Model:
$$y_t = \sigma_t \varepsilon_t,$$
$$\sigma_t^2 = \omega + \alpha(\sigma_{t-1}^2 \varepsilon_{t-1}^2) + \beta \sigma_{t-1}^2 + \gamma_i \tilde{\sigma}_{t-1,i}^2,$$

where the volatility processes of the SV and GARCH classes of models are driven by daily returns with $y_t = R_t$ and $\varepsilon_t \sim \text{NID}(0, 1)$. The SVX and GX models then contain additional intraday information in the variance definition itself with i either equal to 1 or 2, so the realised volatility measures we use in these models are $\tilde{\sigma}_{t,1}$ and $\tilde{\sigma}_{t,2}$ as given in equations (7.4) and (7.5), respectively. The realised volatility models UC-RV and ARFIMA-RV, on the other hand, directly model the realised volatility measure $\tilde{\sigma}_{t,3}$ as defined in equation (7.6).

Table 7.3 presents the estimation results for the above specified models over the full sample period 6 January 1997 to 29 December 2000. The fractional integration parameter estimate in the ARFIMA-RV model has a value of 0.446, exceeding the d estimates found by Ebens (1999) and Andersen *et al.* (2001b) who report values in the region of 0.4. It should

[15]The data series and the programs used for the estimation of the UC-RV, SV and SVX models can be downloaded from the Internte at `www.ssfpack.com/holbook`; also see Appendix C.

be noted that these studies either employ different ARFIMA model specifications or different estimation methods[16]. More importantly however, the logarithmic volatility process might not be covariance-stationary as the hypothesis of $d > 0.5$ cannot be rejected at the 5% significance level[17]. For the UC-RV model we observe estimated values for the first-order autoregressive and moving average parameters of 0.747 and −0.160, respectively, which are both statistically significant. The first-order autocorrelation function of the observations can then be calculated as

$$\hat{\rho}_{(1)} = \frac{\gamma_{\sigma_t^2(1)} + \gamma_{u(1)}}{\gamma_{\sigma_t^2(0)} + \gamma_{u(0)}} = \frac{\sigma_\eta^2(1+\phi\theta)(\phi+\theta)/(1-\phi^2)}{\sigma_\eta^2(1+\theta^2+2\theta\phi)/(1-\phi^2)+\sigma_u^2} = 0.218,$$

as the two components are uncorrelated with each other[18]. The thus fitted autocorrelations for the $\tilde{\sigma}_{t,3}^2$ series are considerably lower than the empirical sample correlations reported in table 7.2 which is mainly due to the fact that we assume $\mathrm{E}(\sigma_u^2) = 1$.

Volatility persistence estimates for the standard SV and the GARCH models are statistically significant and close to unity with $\hat{\phi} = 0.946$ and $\hat{\alpha}+\hat{\beta}$ equal to 0.938 which confirms earlier findings in the literature with regard to daily stock index return series. The SVX and GX models with realised volatility incorporated in the volatility equations show highly significant estimates for the γ regression parameters with $\hat{\gamma}_2$ slightly higher than $\hat{\gamma}_1$. This is to be expected as $\tilde{\sigma}_{t,1}$ is equal to $\tilde{\sigma}_{t,2}$ plus the squared overnight return at time t. In addition we observe that the subsequent log-likelihood values increase compared to those of the SV and the GARCH models. The likelihood ratio statistics for the null hypotheses $\gamma_1 = 0$ and $\gamma_2 = 0$ are 39.44 and 43.14 for the SV and 76.34 and 83.76 for the GARCH models, respectively. They clearly indicate that inclusion of the realised volatility measures $\tilde{\sigma}_{t,1}$ or $\tilde{\sigma}_{t,2}$ significantly improves the fit of the models[19]. With regards to the other parameters we find that the estimates for α in the GX models are negative and no longer statistically significant but that the estimates for σ_η^2 in the SVX models have increased considerably in value. This suggests that the

[16]Ebens (1999), for example, never estimates the autoregressive parameter in his ARFIMAX(p,d,q) model and Andersen *et al.* (2001b) fix the value of the d parameter prior to the estimation of the other parameters in the ARFIMA model.

[17]Also see Oomen (2001) who encounters the same problem with regard to a ten-year sample of FTSE-100 index returns.

[18]See Harvey (1993), pp.30-32. The fitted autocorrelation function for $\tau \geq 2$ is given by $\hat{\rho}_{(\tau)} = \phi\gamma_{\sigma_t^2(\tau-1)}/(\gamma_{\sigma_t^2(0)}+1)$.

[19]Unlike Martens (2002), we find a slightly better in-sample fit when the overnight returns are excluded from the realised volatility measure.

Table 7.3. In-sample estimation results for the Standard & Poor 100 stock index over the period 6 January 1997 to 29 December 2000.

Model Series	UC-RV $\tilde{\sigma}^2_{t,3}$	ARFIMA-RV $\ln \tilde{\sigma}^2_{t,3}$		SV Models y_t				GARCH Models y_t		
μ	1.357 1.149 1.564	0.013 −1.218 1.244	σ^{*2}	1.358 1.073 1.719	1.305 1.169 1.457	1.360 1.217 1.521	ω	0.111 0.019 0.202	0.280 0.015 0.546	0.263 −0.007 0.532
ϕ	0.747 0.662 0.833	−0.061 −0.151 0.029	ϕ	0.946 0.876 0.978	0.360 0.005 0.609	0.357 0.014 0.600	α	0.106 0.056 0.155	−0.022 −0.068 0.025	−0.029 −0.073 0.015
θ	−0.160 −0.317 −0.003		σ^2_η	0.036 0.015 0.087	0.150 0.060 0.379	0.144 0.055 0.374	β	0.832 0.744 0.920	0.286 −0.033 0.605	0.288 −0.034 0.611
σ^2_η	0.279 0.202 0.386		γ_1		0.443 0.347 0.538		γ_1		0.680 0.416 0.944	
σ^2_u		0.286	γ_2			0.465 0.369 0.561	γ_2			0.743 0.468 1.017
d		0.446 0.381 0.512								
$\ln L$	−1581.24	−797.11		−1623.57	−1603.85	−1602.00		−1640.19	−1602.02	−1598.31
$LR(\gamma_1 = 0)$					39.44				76.34	
$LR(\gamma_2 = 0)$						43.14				83.76
AIC	3170.48	1602.21		3253.14	3215.70	3212.00		3286.38	3212.03	3204.61
$Q(12)$	13.528	11.607		21.265	22.794	22.634		20.033	21.405	21.300
N	21.641	32.163		2.549	6.305	5.994		223.26	31.909	26.780

Parameter estimates are reported together with the asymptotic 95% confidence intervals which are asymmetric for σ^{*2}, ϕ in the SV models, and σ^2_η; $LR(\gamma_1 = 0)$ and $LR(\gamma_2 = 0)$ are the likelihood ratio statistics for the hypotheses $\gamma_1 = 0$ and $\gamma_2 = 0$, respectively. AIC is the Akaike Information Criterion which is calculated as -2(ln L) + $2p$; $Q(\ell)$ is the Box-Ljung portmanteau statistic for the estimated observation errors which is asymptotically χ^2 distributed with $\ell - p$ degrees of freedom where p is the total number of estimated parameters; N is the χ^2 normality test statistic with 2 degrees of freedom.

SV type models still benefit from inclusion of both daily and 5-minute returns but that for GARCH models there is little to no incremental value in the daily returns once the information contained in the intraday (and overnight) returns is included[20]. Although we cannot compare the SV and GARCH models in terms of goodness-of-fit, we can confirm that the distributional assumptions with regard to the error term ε_t are more closely followed by the SV than by the GARCH models.

7.6.2 Out-of-Sample Results

Out-of-sample volatility forecasts are constructed from the SV, GARCH and realised volatility models for the evaluation period 13 March 2000 to 29 December 2000 and forecasting horizons range from one day to one week[21]. Tables 7.4 and 7.5 present the forecasting performance results of these models with scaled intraday volatility in equation (7.6) as the measure of realised volatility and forecasts evaluated by means of the regression in equation (7.21) and the error statistics in equations (7.22) and (7.23).

The regression-based results in table 7.4 show that those models which use intraday information produce more accurate out-of-sample volatility forecasts than those that do not. The SV and the GARCH(1,1) model have the lowest coefficients of determination R^2 and as these daily models depend solely on closing prices, extreme price movements during the trading day do not necessarily show up.

For illustrative purposes we plot in figure 7.3 the one-day ahead volatility forecasts of all six models for the first 60 trading days of our out-of-sample forecasting period during which three so-called high volatility days can be observed. The first of these occurs on 4 April 2000 (out-of-sample day 17) but is not recognised as such by the daily volatility models in graph 7.3-ii because daily closing prices only showed a relatively moderate drop of 1%. The other four models incorporate the intraday returns and their volatility forecasts therefore increase after 4 April as can be seen in graph 7.3-i for the realised volatility models and in graph 7.3-iii for the SVX and GX models. The GX model then reacts the most strongly of the four. The second higher volatility period, on the

---

[20]These GARCH model findings confirm the empirical results of Blair *et al.* (2001) who examine Standard & Poor's 100 stock index returns over the earlier 1987 to 1992 period and find values for γ similar to ours. In contrast, Martens (2002) reports on much smaller and statistically insignificant γ estimates for returns on Standard & Poor's 500 futures. Also see: Taylor and Xu (1997).

[21]The SVX and GX models are based on the intraday volatility measure of equation (7.5). Adding the squared overnight returns to the intraday volatility measure produces very similar volatility forecasts although they are slightly less accurate.

Table 7.4. OLS regression out-of-sample forecasting results evaluated against $\tilde{\sigma}^2_{(T+1,T+N),3}$ based on the Standard & Poor's 100 6 January 1997 to 28 December 2000 sample and for the evaluation period 13 March to 29 December 2000

Forecasting Model		Forecasting Horizon $N=1$	$N=2$	$N=3$	$N=4$	$N=5$
UC-RV Model	a	0.5173 (1.877)	1.5421 (2.777)	2.7216 (3.076)	3.8066 (3.132)	4.8179 (3.082)
	b	0.7727 (1.263)	0.6312 (2.250)	0.5579 (2.707)	0.5377 (2.889)	0.5370 (2.881)
	R^2	0.2140	0.1573	0.1274	0.1187	0.1157
ARFIMA-RV Model	a	−0.0199 (0.585)	0.0194 (0.032)	0.3667 (0.414)	0.7123 (0.638)	1.0526 (0.769)
	b	1.1677 (0.679)	1.0498 (0.220)	0.9890 (0.051)	0.9604 (0.196)	0.9461 (0.279)
	R^2	0.2579	<u>0.2352</u>	<u>0.2266</u>	<u>0.2279</u>	<u>0.2310</u>
SV Model	a	0.4627 (1.073)	1.4650 (2.075)	2.5123 (2.251)	3.7111 (2.202)	4.7429 (2.088)
	b	0.7680 (0.864)	0.6181 (2.099)	0.5623 (2.723)	0.5143 (2.855)	0.5059 (2.690)
	R^2	0.0732	0.0580	0.0546	0.0497	0.0501
SVX Model	a	−0.1991 (0.597)	0.4693 (0.929)	1.4373 (2.011)	2.5186 (2.671)	3.7026 (2.998)
	b	1.1385 (0.597)	0.8925 (0.614)	0.7553 (1.648)	0.6709 (2.623)	0.6101 (3.349)
	R^2	<u>0.2705</u>	0.2195	0.1934	0.1802	0.1699
GARCH(1,1) Model	a	0.5868 (1.419)	1.7942 (2.744)	3.0419 (2.991)	4.4492 (2.958)	5.7337 (2.834)
	b	0.6346 (1.586)	0.4776 (3.583)	0.4204 (4.778)	0.3722 (5.109)	0.3570 (4.841)
	R^2	0.0894	0.0606	0.0524	0.0439	0.0415
GX Model	a	0.5453 (2.221)	1.6964 (3.767)	3.0473 (4.398)	4.3803 (4.687)	5.7157 (4.880)
	b	0.6061 (2.889)	0.4614 (4.969)	0.3825 (6.362)	0.3450 (7.482)	0.3227 (8.285)
	R^2	0.2323	0.1778	0.1503	0.1444	0.1442

Parameter estimates and goodness-of-fit R^2 statistics for the OLS regressions as defined in equation (7.21). The t-statistics testing for the null hypotheses $a = 0$ and $b = 1$ are in parentheses and based on standard errors using Newey-West heteroskedasticity and autocorrelation consistent covariance estimates. The highest values for R^2 are underlined.

Table 7.5. Error statistic out-of-sample forecasting results evaluated against $\tilde{\sigma}^2_{(T+1,T+N),3}$ based on the Standard & Poor's 100 6 January 1997 to 28 December 2000 sample and for the evaluation period 13 March to 29 December 2000

Forecasting Model		Forecasting Horizon $N=1$	$N=2$	$N=3$	$N=4$	$N=5$
UC-RV Model	HRMSE	0.6546	0.6326	0.6394	0.6108	0.6040
		2	3	3	3	3
	HMAE	0.4853	0.4572	0.4606	0.4488	0.4511
		2	2	3	3	3
ARFIMA-RV Model	HRMSE	0.6401	0.5613	0.5383	0.5144	0.5058
		1	1	1	1	1
	HMAE	0.4807	0.4406	0.4264	0.4146	0.4100
		1	1	1	1	1
SV Model	HRMSE	0.8583	0.7503	0.7070	0.6823	0.6648
		4	4	4	4	4
	HMAE	0.6561	0.5989	0.5614	0.5402	0.5232
		5	5	4	4	4
SVX Model	HRMSE	0.6684	0.5984	0.5842	0.5555	0.5432
		3	2	2	2	2
	HMAE	0.5134	0.4699	0.4530	0.4346	0.4238
		3	3	2	2	2
GARCH(1,1) Model	HRMSE	1.0093	0.8835	0.8374	0.8154	0.7999
		6	6	5	5	6
	HMAE	0.7600	0.6997	0.6576	0.6331	0.6166
		6	6	6	6	6
GX Model	HRMSE	0.9031	0.8659	0.8787	0.8248	0.7954
		5	5	6	6	5
	HMAE	0.6446	0.5951	0.5820	0.5552	0.5419
		4	4	5	5	5

Error statistics HRMSE and HMAE as defined in equations (7.22) and (7.23), respectively, together with model rankings for the relevant forecasting horizon below these error statistics.

other hand, which is that of 14 to 17 April 2000 (out-of-sample days 25 and 26) also leads to higher volatility forecasts for the SV and GARCH model as daily returns do change considerably for this period. On the whole it appears from the graphs in figure 7.3 that the volatility forecasts of the models with intraday return information follow the realised volatility measure much more closely than those without, confirming the regression based results of table 7.4. More specifically, the goodness-of-

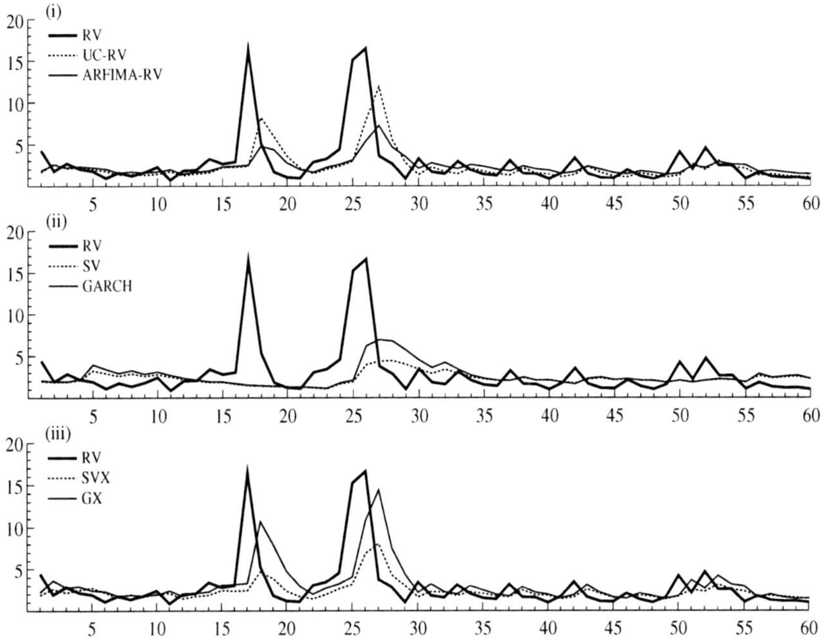

Figure 7.3. One-day ahead volatility forecasts of the (i) UC-RV and ARFIMA-RV, (ii) SV and GARCH, and (iii) SVX and GX models against the realised volatility measure $\tilde{\sigma}^2_{t,3}$ (RV) over the period 13/03/00 to 06/06/00.

fit statistics indicate that the SVX model gives the most accurate one trading day ahead volatility forecasts, whereas the ARFIMA model outperforms the other five models for forecasting horizons $N \geq 2$. Further it has to be noted that the forecasts of the SVX and the ARFIMA-RV model are the least biased for $N = 1$ and $N \geq 2$, respectively. What is more, the hypotheses that the parameter estimates for a and b are equal to 0 and 1, respectively, can never be rejected at the 5% significance level for the ARFIMA-RV model. The worst performing model in this respect is the GX model which has t-statistics that always exceed the critical 5% value of 1.96.

Results change marginally when we evaluate the volatility forecasts against the error statistics in table 7.5. In terms of the HRMSE and HMAE statistics the ARFIMA-RV model now consistently appears to have the most accurate out-of-sample volatility forecasts, followed by the UC-RV for the shorter and the SVX model for the longer forecasting horizons. The worst performing volatility models however are the

GARCH models and even though the SV model does not include the intraday volatility measure it frequently outperforms the GX model. The reason for this appears twofold. Firstly, GARCH models react very strongly to sharp increases in volatility at time T which leads to the overestimation of volatility at time $T+N$ as can be seen in figure 7.3 for the GX model with $N=1$. Secondly, GARCH models appear to have a volatility level which is in general too high[22]. The problem of overestimation is then exacerbated by the choice of error statistics in table 7.5 as overestimation is penalised more severely by the heteroskedasticity adjusted error statistics then underestimation. The value for the ratio that appears in equations (7.22) and (7.23) lies between zero and one when $\mathrm{E}(\sigma^2_{T+1,T+N|T}) < \tilde{\sigma}^2_{(T+1,T+N),3}$, whereas the value of this term is upwardly unbounded when $\mathrm{E}(\sigma^2_{T+1,T+N|T}) > \tilde{\sigma}^2_{(T+1,T+N),3}$.[23]

We therefore conclude that a relatively simple ARFIMA(1,d,0) model for intraday volatility in logarithmic form provides more accurate out-of-sample forecasts than other, more elaborate, volatility models even though the UC-RV and the SVX model also perform very well. When intraday data is not available, the SV model is the preferred volatility model as GARCH models perform worse in terms of error statistics due to overestimation .

7.7 Summary and Conclusions

In this chapter we examine the predictive abilities of six volatility models which we evaluate on the basis of a realised volatility measure that is defined as the scaled sum of squared intraday returns. The models we consider can be divided into realised volatility and daily time-varying volatility models. To the first group belong the Unobserved Components (UC-RV) and the Autoregressive Fractionally Integrated Moving Average (ARFIMA-RV) models where intraday volatility is modelled directly. The Stochastic Volatility (SV) and the Generalised Autoregressive Conditional Heteroskedasticity (GARCH) models, together with their intraday extensions, are defined as daily time-varying models. We empirically investigate the out-of-sample forecasting performance of the various methods for the Standard & Poor's 100 stock index over the period 13 March to 29 December 2000 and for forecasting horizons ranging from one day to one week. We conclude that those models which include

[22] Also see: Andersen *et al.* (2001b) who make the observation that daily GARCH models are not very good at assessing the current volatility level.

[23] We also calculated the RMSE and MAE error statistics which react symmetrically to under- and overestimations; in terms of model ranking little changed but relative differences between error statistics decreased considerably.

the intraday information perform better than those that are solely based on daily returns. The most accurate forecasts are then obtained with the ARFIMA-RV model, followed by the SVX and the UC-RV model. Although the GARCH model extended with intraday volatility appears to perform well when its forecasts are evaluated on the basis of regression methods, other evaluation criteria indicate that it tends to overestimate volatility. As the GARCH(1,1) model also suffers from this problem we conclude that in the absence of intraday volatility information the SV model is the preferred model for forecasting volatility.

Chapter 8

CONCLUSIONS

Volatility plays an important role in financial decision making which has motivated extensive research into the modelling of volatility processes. Thus far most empirical studies on stock market volatility have focused on the Generalised Autoregressive Conditional Heteroskedasticity (GARCH) model and its extensions which have proved to be very successful. In this book we have investigated and analysed the variability of stock index returns in the context of Stochastic Volatility (SV) models which have their origin in theoretical finance and represent an appealing alternative to the deterministic GARCH models.

After the introduction in Chapter 1 in which our research objectives were motivated and where an overview of the book was provided, Chapter 2 discussed the empirical stylised facts of stock index return series at different frequencies. Using the Financial Times All Share Index for illustrative purposes we confirmed the existence of volatility persistence, which proved to be especially pronounced for the higher frequency series, and observed that the return distributions were leptokurtic compared to the normal distribution. We proceeded with the discussion of the model specifications and the basic properties of the two classes of time-varying volatility models. Whereas GARCH models define the volatility process as a function of past observations and variances, SV models specify it as an unobserved component that follows some stochastic process. Both GARCH and SV models explained the high degree of autocorrelation in the squared returns and the excess kurtosis in the returns. These time-varying volatility models therefore appeared promising for the description and forecasting of asset return volatility. The main difference was that GARCH models were already widely regarded as very useful volatility models that had proven their worth in many empirical studies and that research into SV models was limited due to the fact that these models were more difficult to estimate. If however SV models were to outperform the GARCH models, the additional effort would be justified. For this purpose SV and GARCH model specifications were fitted to daily and weekly returns series of the Financial Times All Share Index. We found that the GARCH model with normally distributed error terms was the worst performing model and that the t-distributed GARCH model appeared to give a slightly better in-sample fit than the Gaussian SV model although at the cost of an additional model parameter. Theoretical sample autocorrelation results deteriorated however when leptokurtic error distributions were imposed and we therefore concluded that the SV model best captured the salient features of stock market return volatility.

Chapter 3 investigated the relationship between stock market returns and their volatility in the context of SV models for three international stock indices. Although a positive relation between expected excess returns and expected volatility was both intuitively appealing and consistent with financial theory, empirical research had been unable to establish a convincing positive relationship using GARCH-M models. Instead there appeared to be stronger evidence of a negative relation between unexpected returns and innovations to the volatility process which French *et al.* (1987) interpreted as indirect evidence of a positive correlation between the expected risk premium and *ex ante* volatility as this was one of the assumptions underlying their so-called volatility feedback theory. With the Stochastic Volatility in Mean (SVM) model we found that the in-mean parameter estimate d was negative for all series, although the hypothesis of d equal to zero could never be rejected at the conventional 5% significance level. Interestingly, the estimates for the d parameter in the GARCH-M model were all positive though not significant. The difference in the sign of the in-mean parameter was attributable to the fact that SV models, by construction, contained an error term in the variance equation which entered directly into the mean equation of the SVM model. As the shock to the volatility process should not be related to the predictable components of the return process, d in the SVM model therefore not only measured the *ex ante* relation between excess returns and volatility but also the relation between the *unexpected* components. The in-mean estimate in the deterministic GARCH-M model, on the other hand, only measured the *ex ante* relationship between excess returns and volatility. We therefore concluded that the negative relation between the unexpected components in the SVM model dominated the weaker presumably positive relation between the expected components, mirroring earlier finding by French *et al.* (1987). The largest values for the d parameter estimate in both the SVM model and the GARCH-M model, albeit with opposite signs, were observed for the shorter samples which started after the stock market crash of 1987. This was interpreted as confirmation of our hypothesis that a more positive *ex ante* relationship between the return and volatility process in the post-crash sample induced an even stronger volatility feedback effect which dominated the relation between expected returns and expected volatility even more convincingly thus forcing the d estimate in the SVM model even further downward.

In Chapter 4 the forecasting performance of a variety of volatility models was discussed. Empirical research in this area had resulted in contradictory findings and we therefore investigated the issue with four different models and for six international stock indices. We constructed daily volatility forecasts for returns on the UK Financial Times All Share, US Standard & Poor's Composite, Japanese Topix, German DAX, French CAC 40 and the Dutch AEX Index and the models we selected were the SV, GARCH, Random Walk (RW) and the Exponentially Weighted Moving Average (EWMA) model. With forecasting horizons N of 1 to 20 days we updated the estimation samples on a daily basis as for such small values for N volatility forecasts of the SV and GARCH models were shown to be mainly determined by short-term volatility. We then found that the SV and GARCH models produced the most accurate forecasts but that their ranking appeared to be sensitive to the choice of evaluation criteria. As our results favoured the SV to the GARCH model we concluded that daily stock market volatility was best predicted with the SV models. This confirmed the findings by Heynen (1995) and Yu (2000) who thus far had been the only ones to investigate the forecasting performance of SV models in the context of stock markets.

The next chapter described the concept of implied volatility. The early literature supported the hypothesis that implied volatility provided a more accurate forecast of future stock market volatility than historical volatility and that the most accurate implied volatility forecasts were inferred from at or near-the-money options. Day and Lewis (1992) then conducted the first comparative study that employed GARCH models and although volatility forecasts based on GARCH models outperformed the implied volatility measure, the best forecasts were obtained when both sources of information were used. Subsequent research in the early nineties produced conflicting findings and the assumed comprehensive information content of implied volatility appeared questionable. However, more recent empirical evidence strongly suggests that historical data, even when used in combination with GARCH models, contains little or no incremental information about future volatility.

Chapter 6 examined the predictive abilities of SV models for daily returns on the Standard & Poor's 100 stock index. Their forecasting performance was compared with that of implied volatility for which the VIX index of the Chicago Board Options Exchange (CBOE) was used. The VIX is constructed in such as way that many of the measurement errors which might induce biased implied volatility measures are mitigated but we still expected the VIX to be upwardly biased as, for example, the wildcard option implicit in OEX options was ignored. We developed an SV model with the implied volatility measure embedded in the variance equation and termed this the Stochastic Volatility with eXogenous variables (SVX) model. Such a model definition induced a multiple lagged implied volatility structure and hence we also constructed an SVX model with persistence adjustment term which was referred to as the SVX$^+$ model. The Stochastic Implied Volatility (SIV) model was obtained by restricting the ϕ parameter in the SVX models to equal zero which allowed for statistical tests for nested models. The in-sample results indicated that the SV class models which included implied volatility measures such as the VIX index produced more effective estimates of the underlying volatility process of stock index returns than dynamic models solely based on historical returns. The out-of-sample volatility forecasts were constructed for forecasting horizons ranging from 1 to 20 trading days and realised volatility was approximated both as daily squared returns and as cumulative intraday squared returns following research by, for example, Andersen and Bollerslev (1998). For the intraday volatility measure a sampling frequency of 10 minutes was selected. The relative forecasting accuracy of the various volatility models was evaluated using both regression-based evaluation methods and error statistics. The thus obtained results showed that the SIV model outperformed both the SV and the SVX$^+$ model but that the forecasts of the SV model were the least biased. We found that all models on average underestimated the volatility process which was attributable to the fact that the average level of the estimation samples was much lower than that of the evaluation period; this lower level resulted in smaller scaling parameter estimates. Even though the VIX itself appeared to overestimate realised volatility, it was shown that embedding VIX into the variance equation of the SV models resulted in underestimation with volatility forecasts that were on average lower than those of the standard SV model.

Chapter 7 evaluated the forecasting performance of a wide variety of volatility models for the Standard & Poor's 100 stock index series using intraday data. The forecasting performance of these models was evaluated on the basis of a realised volatility measure which was defined as the scaled sum of squared intraday returns. The volatility models were divided into two groups: To the first group belonged the so-called realised volatility models which modelled intraday volatility directly. The Unobserved Components (UC-RV) model developed in this chapter was a discrete-time approximation of the continuous time series process that consisted of independent Ornstein Uhlenbeck (OU) processes and resulted in an unobserved components model with independent ARMA components and an error term. The forecasting performance of this type of volatility model had not been researched before. The in logarithms defined ARFIMA-RV model, on the other hand, had already been successfully applied in the context of volatility forecasting by, for example, Andersen *et al.* (2001b). The second group of volatility models consisted of the Stochastic Volatility and the GARCH models together with their intraday extensions; these daily time-varying volatility models defined the volatility process as the second moment of daily returns. Although we also reported on in-sample findings, the emphasis in this chapter was on the out-of-sample forecasting results. Volatility forecasts were constructed for forecasting horizons ranging from one day to one week for each of the six models and the evaluation criteria included both regression-based and heteroskedasticity adjusted error statistics. On the basis of these we concluded that those models which contained intraday information performed better than those which only used daily returns. The most accurate forecasts were then obtained with the ARFIMA-RV model, followed by the SV model with embedded intraday volatility and the UC-RV model. The GARCH model extended with intraday volatility also appeared to perform well when its forecasts were evaluated on regression-based methods but other evaluation criteria indicated that it tended to overestimate volatility. As the daily GARCH model suffered from the same problem we concluded that in the absence of intraday volatility information the SV model was the preferred model.

In this book some interesting extensions of Stochastic Volatility models have been presented for the first time and their empirical implications have been investigated and compared with other methods for a number of international stock indices. In order to undertake the various studies the problem of model estimation had to be overcome but in recent years modelling techniques have been developed to resolve these problems. The extra effort proved to be worthwhile as it led to further insights on volatility modelling and improved volatility forecasting performance. The challenge will continue however as the volatility processes of financial time series remains difficult to quantify and there is still much to be learned. A second strand of recent research that looks very promising is that of modelling volatility with high frequency data. We are convinced that further empirical finance research into SV models, as well as the increasing availability of high frequency data, will lead to a better understanding of financial market volatility; research into these new areas has only just begun.

Appendix A
Estimation of the SVM Model

In this appendix we show how the parameters of the SVM model are estimated by simulated maximum likelihood[1].

Further, we show how to compute the conditional mean and variance of the volatility process h_t.

A.1 Model

To simplify the exposition we initially consider the model

$$\begin{aligned} y_t &= d\sigma^{*2}\exp(h_t) + \sigma^* \exp(0.5h_t)\varepsilon_t, \\ h_t &= \phi h_{t-1} + \sigma_\eta \eta_t, \end{aligned} \tag{A.1}$$

where y_t denotes the underlying series of interest, in our case these are stock index returns. The disturbances ε_t and η_t are standard normally distributed and they are mutually and serially uncorrelated. The latent variable h_t is modelled as a stationary Gaussian autoregressive process of order 1 and with $0 < \phi < 1$. The unknown parameters are collected in the vector

$$\psi = (\phi, \sigma_\eta, \sigma^{*2}, d)'.$$

The nature of the model is conditionally Gaussian but we deal with a non-linear model since the variance of the overall disturbance term in y_t is given by $\sigma^{*2}\exp(h_t)$ which is stochastic. The Gaussian density for ε_t can be replaced by other continuous distributions. We note that the conditional Gaussian density function $p(y|\theta, \psi)$ of the SVM model with

[1]The material in this appendix was previously published in the *Journal of Applied Econometrics* (2002) **17**, 667-689, as "The Stochastic Volatility in Mean Model: Empirical evidence from international stock markets" by S.J. Koopman and E. Hol Uspensky. Reprinted with permission of John Wiley & Sons Limited.

$$\theta = (h_1, \ldots, h_T)',$$

is log-concave in h_t. This property is useful when employing importance sampling.

The techniques presented in the sections below for model (A.1) can be adjusted straightforwardly to deal with the full model (3.1), where σ_t^2 is given by equation (3.3) and μ_t is given by (3.6), since the extensions do not interact directly with the stochastic variance σ_t^2. However it will be indicated in the following sections where details are different for the full model.

A.2 Likelihood Evaluation Using Importance Sampling

The construction of the likelihood for the SVM model is complicated because the latent variable h_t appears in both the mean and the variance of the SVM model. We adopt the Monte Carlo likelihood approach developed by Shephard and Pitt (1997) and Durbin and Koopman (1997). This simulation method of computing the loglikelihood function can be derived as follows.

Define the likelihood as

$$L(\psi) = p(y|\psi) = \int p(y, \theta|\psi)d\theta = \int p(y|\theta, \psi)p(\theta|\psi)d\theta. \tag{A.2}$$

An efficient way of evaluating this likelihood is by using Monte Carlo integration and, in particular, importance sampling; see Ripley (1987, Chapter 5). We require a simulation device for sampling from an importance density $\tilde{p}(\theta|y, \psi)$ which relates to the true density $p(\theta|y, \psi)$. We follow standard practice in developing a Gaussian approximating model for y with the assumption of given values for θ and ψ and with the density denoted by $g(y|\theta, \psi)$. The approximating model is constructed in such a way that the first two moments of $p(y|\theta, \psi)$ and $g(y|\theta, \psi)$ are equal. Since $g(y|\theta, \psi)$ is Gaussian, it will be relatively straightforward to sample from $\tilde{p}(\theta|y, \psi) = g(\theta|y, \psi)$. The conditions under which this approximation converges almost surely to the true value as the number of simulations from the importance density increases are given by Geweke (1989, Theorem 1). An approximating Gaussian model for the SVM model is developed in section A.3. Simulation smoothers such as the ones of de Jong and Shephard (1995) and Durbin and Koopman (2002) can be used to sample from the importance density $g(\theta|y, \psi)$ in the case of SV models.

The likelihood function (A.2) is rewritten as

$$L(\psi) = \int p(y|\theta, \psi)\frac{p(\theta|\psi)}{g(\theta|y, \psi)}g(\theta|y, \psi)d\theta = \tilde{\mathrm{E}}\{p(y|\theta, \psi)\frac{p(\theta|\psi)}{g(\theta|y, \psi)}\}, \tag{A.3}$$

where $\tilde{\mathrm{E}}$ denotes expectation with respect to the importance density $g(\theta|y, \psi)$. Expression (A.3) can be simplified using a suggestion of Durbin and Koopman (1997). The likelihood function associated with the importance density is given by

$$L_g(\psi) = g(y|\psi) = \frac{g(y, \theta|\psi)}{g(\theta|y, \psi)} = \frac{g(y|\theta, \psi)p(\theta|\psi)}{g(\theta|y, \psi)}, \tag{A.4}$$

and it follows that

$$\frac{p(\theta|\psi)}{g(\theta|y,\psi)} = \frac{L_g(\psi)}{g(y|\theta,\psi)}.$$

This ratio also appears in (A.3) and substitution leads to

$$L(\psi) = L_g(\psi)\,\tilde{\mathrm{E}}\{\frac{p(y|\theta,\psi)}{g(y|\theta,\psi)}\}, \tag{A.5}$$

which is the convenient expression we will use in our calculations. The likelihood function of the approximating Gaussian model can be calculated via the Kalman filter and the two conditional densities are easy to compute given a value for θ. It follows that the likelihood function of the SVM model is equivalent to the likelihood function of an approximating Gaussian model, multiplied by a correction term. This correction term only needs to be evaluated via simulation.

An obvious estimator for the likelihood of the SVM model is

$$\hat{L}(\psi) = L_g(\psi)\bar{w}, \tag{A.6}$$

where

$$\bar{w} = \frac{1}{M}\sum_{i=1}^{M} w_i, \qquad w_i = \frac{p(y|\theta^i,\psi)}{g(y|\theta^i,\psi)}, \tag{A.7}$$

and θ^i denotes a draw from the importance density $g(\theta|y,\psi)$. The accuracy of this estimator depends on the properties of the so-called weights w_i; see Geweke (1989). Since the simulation samples are independent of each other, it follows immediately that the variance due to simulation decreases as M increases. In practice, we usually work with the log of the likelihood function to manage the magnitude of density values. The log transformation of $\hat{L}(\psi)$ introduces bias for which we can correct up to order $O(M^{-3/2})$; see Shephard and Pitt (1997) and Durbin and Koopman (1997). We obtain

$$\ln \hat{L}(\psi) = \ln L_g(\psi) + \ln \bar{w} + \frac{s_w^2}{2M\bar{w}^2}, \tag{A.8}$$

with $s_w^2 = (M-1)^{-1}\sum_{i=1}^{M}(w_i - \bar{w})^2$.

A.3 Approximating Gaussian Model Used For Importance Sampling

The approximating model is based on a linear Gaussian model with mean $\mathrm{E}(y_t) = h_t + c_t$ and variance $\mathrm{V}(y_t) = H_t$, that is

$$y_t = h_t + u_t, \qquad u_t \sim \mathrm{N}(c_t, H_t), \qquad t = 1, \ldots, n, \tag{A.9}$$

where c_t and H_t are determined in such a way that the mean and variance of y_t implied by the approximating model (A.9) and by the true model, under consideration, (A.1) are as close as possible[2].

We achieve this by equalising the first and second derivatives of $p(y|\theta, \psi)$ and $g(y|\theta, \psi)$ with respect to θ at $\hat{\theta} = \tilde{\mathrm{E}}(\theta) = \int \theta g(\theta|y, \psi)$. Note that $p(\cdot)$ refers to a density for the true model and $g(\cdot)$ refers to a density for the approximating Gaussian model. Further, it follows that $\hat{\theta}$ can simply be obtained via the Kalman filter and smoother applied to the approximating model (A.9). The conditional densities are given by

$$p(y|\theta, \psi) = \prod_{t=1}^{n} p_t, \qquad g(y|\theta, \psi) = \prod_{t=1}^{n} g_t, \tag{A.10}$$

with

$$\begin{aligned} p_t &= p(y_t|h_t, \psi) = -0.5[\ln 2\pi\sigma^2 + h_t + \exp(-h_t)\sigma^{-2}\{y_t - d\exp(h_t)\}^2], \\ g_t &= g(y_t|h_t, \psi) = -0.5\{\ln 2\pi + \ln H_t + H_t^{-1}(y_t - c_t - h_t)^2\}. \end{aligned} \tag{A.11}$$

Differentiating both densities twice with respect to h_t gives

$$\begin{aligned} \dot{p}_t &= -0.5[1 + \sigma^{-2}\{d^2\sigma^{*2}\exp(h_t) - y_t^2\exp(-h_t)\}], \\ \ddot{p}_t &= -0.5\sigma^{-2}[d^2\sigma^{*2}\exp(h_t) + y_t^2\exp(-h_t)], \\ \dot{g}_t &= H_t^{-1}(y_t - c_t - h_t), \\ \ddot{g}_t &= -H_t^{-1}. \end{aligned}$$

Equalising the first and second derivatives, that is $\dot{p}_t = \dot{g}_t$ and $\ddot{p}_t = \ddot{g}_t$ for $t = 1, \ldots, n$, leads to

$$\begin{aligned} c_t &= y_t - h_t + 0.5H_t[1 + \sigma^{-2}\{d^2\sigma^{*2}\exp(h_t) - y_t^2\exp(-h_t)\}], \\ H_t &= 2\sigma^{*2}/[d^2\sigma^{*2}\exp(h_t) + y_t^2\sigma^{*-2}\exp(-h_t)]. \end{aligned}$$

The resulting model for $\tilde{y}_t = y_t - c_t$ is equivalent to

$$\tilde{y}_t = h_t + \tilde{u}_t, \qquad \tilde{u}_t \sim \mathrm{N}(0, H_t), \qquad t = 1, \ldots, n,$$

with

$$\begin{aligned} \tilde{y}_t &= h_t - \frac{\sigma^{*2} + d^2\sigma^{*2}\exp(h_t) - y_t^2\sigma^{*-2}\exp(-h_t)}{d^2\sigma^{*2}\exp(h_t) + y_t^2\sigma^{*-2}\exp(-h_t)}, \\ H_t &= \frac{2\sigma^2}{d^2\sigma^{*2}\exp(h_t) + y_t^2\sigma^{*-2}\exp(-h_t)}. \end{aligned}$$

[2]Note that the true model describes a non-linear relationship between y_t and h_t; the approximating (linear) model is effectively a second-order Taylor expansion of the true model around h_t. Further, the multivariate Gaussian density $g(\theta|y, \psi)$ can be regarded as a Laplace approximation to the true density $p(\theta|y, \psi)$.

It should be noted that $H_t > 0$ for any value of h_t. We cannot solve out for $\tilde{y}_t$ and H_t at $\hat{h}_t = \tilde{\mathrm{E}}(h_t)$ because $\tilde{\mathrm{E}}$ refers to expectation with respect to the approximating model which depend on h_t. However, such complicated but linear system of equations is usually solved iteratively by starting with a trial value $h_t = h_t^*$. Computing $\tilde{y}_t$ and H_t based on h_t^* and applying the Kalman filter smoother to model (A.9) leads to a smoothed estimate for h_t which can be used as a new trial value for h_t. Recomputing $\tilde{y}_t$ and H_t based on this new trial value leads to an iterative procedure which converges to $\hat{h}_t$. Note that the first and second derivatives of the true and approximating densities are equal at $h_t = \hat{h}_t$. More details are given by Durbin and Koopman (1997). It is worth mentioning that $\hat{h}_t$ is equal to the mode of $p(h_t|y, \psi)$ which can be of interest.

When we consider the full model (3.1), with σ_t^2 given by equation (3.3) and μ_t given by (3.6), the likelihood function is only affected in the squared error term. The last term of the definition of p_t in (A.11) is replaced by the term

$$\exp(-h_t)\sigma^{-2}\{y_t - a - by_{t-1} - d\exp(h_t)\}^2.$$

We observe that the extensions do not change the stochastic process for h_t. Therefore, the simulation scheme for computing the Monte Carlo likelihood remains the same. However, the approximating model changes slightly; that is, $\dot{p}_t$ changes but $\ddot{p}_t$ does not change. In other words, the definition for c_t changes but the definition for H_t does not. Finally, numerical maximisation of the Monte Carlo likelihood is now also with respect to the parameters a and b.

A.4 Monte Carlo Evidence of Estimation Procedure

In this section we present some results of a Monte Carlo study which is carried out to investigate the small sample performance of the estimation procedure presented in section A.2. In short, we generate K simulated SVM series for the model presented in section A.1 and for some given 'true' parameter vector ψ. Subsequently, we treat ψ as unknown and estimate it for each series using the maximum likelihood method described in section A.2. We compute the sample mean and standard deviation together with a histogram for each element in ψ and compare it with the 'true' parameter value.

The details of the likelihood estimation procedure are as follows. For a given parameter vector ψ, we obtain the approximating Gaussian model as described in section A.3. The Gaussian loglikelihood function of the approximating model $\ln L_g(\psi)$ can be computed using the standard Kalman filter; see, for example, Durbin and Koopman (2001). We have used the simulation smoothing method of de Jong and Shephard (1995) to generate the importance samples θ^i $(i = 1, \ldots, M)$ but recently a more simpler method has been developed by Durbin and Koopman (2002) which can also be used for this purpose. The polar method as implemented in Ox by Doornik (1998) is used for generating standard normal deviates which are required as input for the simulation smoother to obtain samples θ^i from $g(\theta|y, \psi)$. The random number generator needs to be initialised by some fixed value. The importance samples θ^i allow us to compute weight w_i using (A.7) for $i = 1, \ldots, M$ where M is set equal to 200 for all our calculations. For each sample θ^i, three antithetic variables are computed in the way described by Durbin and Koopman (1997). This leads to a total of 800 weights for each likelihood evaluation. The mean $\bar{w}$ and variance s_w^2 of these weights are then used to compute the importance sampling estimate of the loglikelihood value (A.8). For the purpose of estimating the parameter vector ψ, this likelihood estimation procedure is used repeatedly for different values of vector ψ. It is noted that the random number generator is initialised by the same fixed value for each loglikelihood computation so that the numerical search procedure for the maximum likelihood is not affected by randomness. The BFGS maximisation method is used to maximise the simulated likelihood function with respect to ψ; the BFGS method is documented by, among others, Fletcher (1987) and it is implemented in Ox by Doornik (1998). This estimation procedure worked satisfactorily in our simulation study below as well as in our empirical study of section 3.5. In our implementation, a single likelihood evaluation of an SVM model with 5000 observations took 0.7 seconds on a Pentium III 800 Mhz when employing $M = 200$ importance samples. Maximum likelihood estimation of parameters was on average completed in 20 seconds[3].

Parameter estimation is not with respect to vector ψ as defined in section A.1, but with respect to transformed parameter vector ψ^*. The autoregressive parameter ϕ is restricted to have a value between zero and one; therefore we estimate ψ_1^* where

$$\phi = \psi_1 = \frac{\exp(\psi_1^*)}{1+\exp(\psi_1^*)}, \qquad \psi_1^* = \ln\frac{\phi}{1-\phi}.$$

Further, we estimate the log variance σ^{*2} and the log standard deviation σ_η. The mean parameter d is estimated without transformation.

We start by considering the standard SV model, that is the SVM model of section A.1 with $d = 0$. Therefore, the last element of ψ is omitted. For generating Monte Carlo samples, the 'true' parameter values are set to

	ψ	ψ^*
$\psi_1 = \phi$	0.97	3.5
$\psi_2 = \sigma_\eta$	.135	-2
$\psi_3 = \sigma^{*2}$	.549	-0.3

which are typical values found in our empirical study of section 3.5.

The Monte Carlo results for the basic SV model are similar but slightly better compared to results presented in similar studies of Jacquier *et al.* (1994) and Sandmann and Koopman (1998). Note that in these studies the parameter values were not transformed and that the estimation procedures used were different from ours. The results given in figure A.1 are for the typical sample size $n = 5000$ with the number of iterations set to $K = 500$. The graphical output includes a histogram of the estimated parameter values and an estimated density function which is computed using a standard non-parametric Gaussian kernel method.

The sample mean and standard deviation of the K estimated coefficients are given by

[3]Relevant programs for the estimation of the SVM model can be found on the Internet at `www.ssfpack.com/holbook`; also see Appendix C.

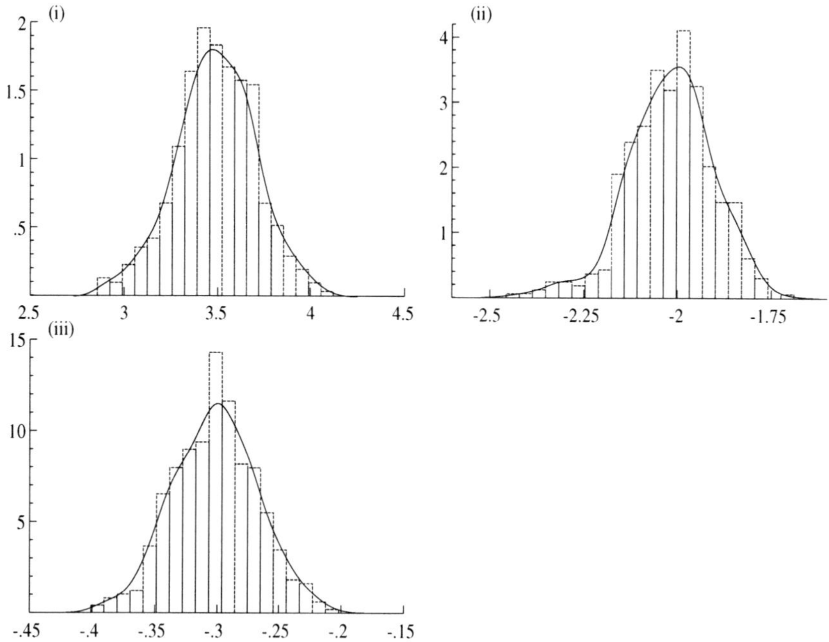

Figure A.1. Monte Carlo results for the standard SV model. Histograms and estimated densities (solid lines) of the maximum likelihood parameter estimates are presented for the SV model with (i) $\psi_1^* = 3.5$, (ii) $\psi_2^* = -2$ and (iii) $\psi_3^* = -0.3$. The Monte Carlo experiment is based on $K = 500$ iterations and sample size $n = 5000$.

	'true'	mean	stand.dev	mean asym stand.err
ψ_1^*	3.5	3.490	0.215	0.225
ψ_2^*	-2.0	-2.020	0.112	0.111
ψ_3^*	-0.3	-0.301	0.0338	0.0337

The last column contains the averages of the asymptotic standard errors of the estimates and they are conveniently close to the sample standard deviations of the estimates. The Monte Carlo results can also be presented in terms of vector ψ; we note that the resulting confidence intervals are asymmetric due to the non-linear transformations. We obtain

	mean	LHS "95% CI"	RHS "95% CI"
$\psi_1 = \phi = 0.97$	0.970	0.955	0.981
$\psi_2 = \sigma_\eta = 0.135$	0.133	0.107	0.165
$\psi_3 = \sigma^{*2} = 0.549$	0.547	0.480	0.625

where LHS is the left-hand-side border and RHS is the right-hand-side border of the 95% confidence interval. These results will be used as a benchmark for the Monte Carlo results of the SVM model.

We now turn our attention towards the Monte Carlo evidence for the SVM model. We keep the 'true' parameters of the SV model and look at the results for a typical value of d, that is $d = 0.1$. The Monte Carlo experiments are again based on $n = 5000$ and $K = 500$. The results for the SVM model with 'true' $d = 0.1$ are given in figure A.2 and the sample statistics are given by

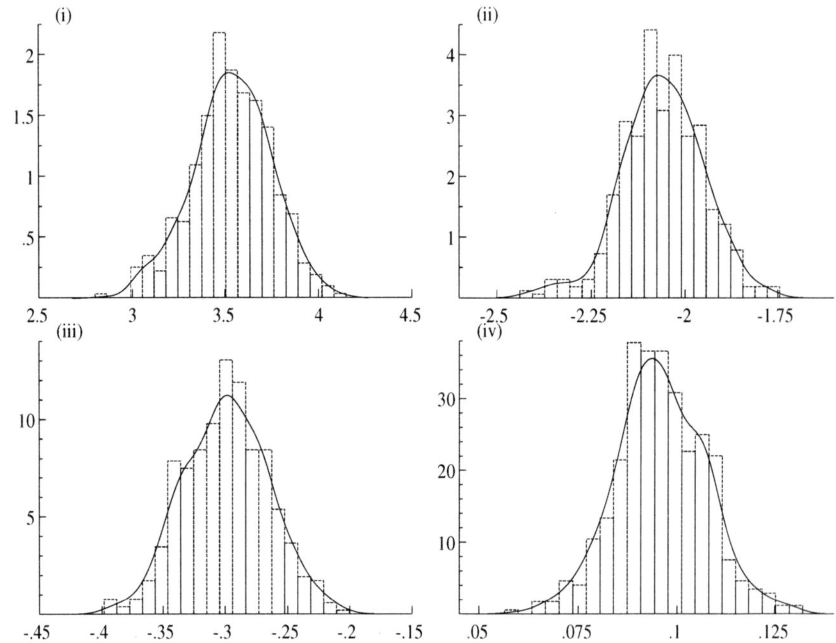

Figure A.2. Monte Carlo results for the SVM model.
Histograms and estimated densities (solid lines) of the maximum likelihood parameter estimates are presented for the SVM model with (i) $\psi_1^* = 3.5$, (ii) $\psi_2^* = -2$, (iii) $\psi_3^* = -0.3$ and (iv) $\psi_4^* = 0.1$. The Monte Carlo experiment is based on $K = 500$ iterations and sample size $n = 5000$.

	'true'	mean	stand.dev	mean asym stand.err
ψ_1^*	3.5	3.537	0.210	0.213
ψ_2^*	−2.0	−2.059	0.108	0.103
ψ_3^*	−0.3	−0.299	0.0341	0.0338
ψ_4^*	0.1	0.0960	0.0113	0.0115

The standard deviation of the estimates for ψ^* obtained from the Monte Carlo samples are very close to the averages of the asymptotic standard errors of the estimates. This indicates that asymptotic standard errors can be used for estimates obtained from the methods of section A.2. The results in terms of ψ are given by

	mean	LHS "95% CI"	RHS "95% CI"
$\psi_1 = \phi = 0.97$	0.972	0.958	0.981
$\psi_2 = \sigma_\eta = 0.135$	0.128	0.104	0.156
$\psi_3 = \sigma^{*2} = 0.549$	0.550	0.481	0.628
$\psi_4 = d = 0.1$	0.0960	0.0733	0.119

Comparing the results for the standard SV model, we conclude that the confidence intervals for ψ_1, ψ_2 and ψ_3 are very similar and that the in-mean parameter d can be accurately estimated with a relatively small standard deviation[4].

Finally, we have assessed the variability of the importance sampling estimator due to the choice of different random numbers. This variability depends on the number of simulations M in (A.7) and by setting this value to 200 we obtained results which only varied marginally when different random numbers were used. By repeating the Monte Carlo experiment for different values of n we found that the variability of the estimator increases when n gets larger but not dramatically.

[4]We repeated this experiment with various values for d and the outcome only changed marginally. Similar overall results were obtained for simulations based on the smaller sample size $n = 500$; these can be found in Koopman and Hol Uspensky (2000).

Appendix B
Estimation of the SVX Models

In this appendix we discuss how the parameters of the SVX class of models are estimated by simulated maximum likelihood using importance sampling techniques[1].

B.1 The SVX Model in State Space Form

Let $y = (y_1, \ldots, y_T)'$ and $\theta = (\theta_1, \ldots, \theta_T)'$ where observation y_t is modelled as in equation (6.1) and its log-volatility is given by

$$\sigma_t^2 = \exp(\theta_t), \qquad \theta_t = \gamma^* + h_t,$$

with signal h_t modelled as in equations (6.3), (6.4) or (6.5), for $t = 1, \ldots, T$. Note that $\sigma^{*2} = \exp(\gamma^*)$ in equation (6.2). Further we shall collect the parameters which are not included in the state vector below in the parameter vector ψ. The SVX model (6.1), (6.2) and (6.4) in state space form is given by

$$p(y|\theta, \psi) = \prod_{t=1}^{T} \mathrm{N}(0, \sigma_t^2),$$

with $\mu_t = 0$ and $\sigma_t^2 = \exp(\theta_t) = \exp(\gamma^* + h_t) = \sigma^{*2}\exp(h_t)$. The state vector collects the components of the log-volatility and is given by $\alpha_t = (\gamma^*, \gamma, h_t)'$. The so-called transition equation for the state vector is given by

$$\alpha_{t+1} = \begin{bmatrix} 1 & 0 & 0 \\ 0 & 1 & 0 \\ 0 & x_t & \phi \end{bmatrix} \alpha_t + \begin{pmatrix} 0 \\ 0 \\ \sigma_\eta \end{pmatrix} \eta_t,$$

where disturbances η_t are distributed as $\mathrm{NID}(0, 1)$. The initial state vector α_1 is given by

[1]Estimation methods for the SVX class of models were developed together with Siem Jan Koopman; see Hol and Koopman (2000).

$$\begin{pmatrix} \gamma^* \\ \gamma \\ h_1 \end{pmatrix} \sim \mathrm{N}\left\{ \begin{pmatrix} 0 \\ 0 \\ 0 \end{pmatrix}, \begin{bmatrix} \kappa & 0 & 0 \\ 0 & \kappa & 0 \\ 0 & 0 & \sigma_\eta^2/(1-\phi^2) \end{bmatrix} \right\},$$

for some arbitrary large value for κ. This specification follows since γ^* and γ are unknown while the process of h_t is stationary. The variance of h_1 equals the unconditional variance of the autoregressive process of h_t. Finally, the parameter vector ψ is given by

$$\psi = \begin{pmatrix} \phi \\ \sigma_\eta \\ \sigma_\epsilon \end{pmatrix}.$$

This completes the model specification in state space form. The density for y given θ is given by

$$p(y|\theta,\psi) = \prod_{t=1}^{T} \mathrm{N}(0, \exp\theta_t), \qquad \theta_t = (1,0,1)\alpha_t.$$

Given a realisation of θ it follows that this density can be computed straightforwardly. This representation of SV and SVX models can be taken as a special case of the non-linear state space model.

B.2 Parameter Estimation by Simulated Maximum Likelihood

We will adopt the Monte Carlo likelihood approach of Shephard and Pitt (1997) and Durbin and Koopman (1997) for the construction of the exact likelihood function of the SV model. This method can be modified straightforwardly for the SVX class of models since the non-linear relation between the log-volatility h_t and the observation equation is not altered in the SVX case; only the specification for h_t is different. Similar considerations are discussed by Chib, Nardari and Shephard (1998) in a Bayesian context using Markov chain Monte Carlo methods. The same modification can be used for the SVX$^+$ model since we merely replace the explanatory variable x_{t-1} by $(1-\phi L)x_{t-1}$.

The loglikelihood function for the SVX model can be computed via the Monte Carlo technique of importance sampling. The likelihood function can be expressed as

$$L(\psi) = p(y|\psi) = \int p(y,\theta|\psi)d\theta = \int p(y|\theta,\psi)p(\theta|\psi)d\theta. \tag{B.1}$$

An efficient way of evaluating the likelihood is by using importance sampling; see Ripley (1987, Chapter 5). We require a simulation device to sample from an importance density $\tilde{p}(\theta|y,\psi)$ which we prefer to be close to the true density $p(\theta|y,\psi)$. An obvious choice for the importance density is the conditional Gaussian density since in this case it is relatively straightforward to sample from $\tilde{p}(\theta|y,\psi) = g(\theta|y,\psi)$. An approximating Gaussian model for the SVX model is obtained by equalising the first and second derivatives of $p(y|\theta)$ and $g(y|\theta)$ with respect to θ. A simulation smoother is used to sample from the approximating Gaussian model $g(\theta|y,\psi)$; a simple and efficient procedure for this is developed by Durbin and Koopman (2002).

The likelihood function (B.1) can be obtained by writing

$$L(\psi) = \int p(y|\theta,\psi)\frac{p(\theta|\psi)}{g(\theta|y,\psi)}g(\theta|y,\psi)d\theta = \tilde{\mathrm{E}}\{p(y|\theta,\psi)\frac{p(\theta|\psi)}{g(\theta|y,\psi)}\}, \tag{B.2}$$

where $\tilde{\mathrm{E}}$ denotes expectation with respect to the importance density $g(\theta|y,\psi)$. The likelihood function of the approximating Gaussian model is given by

$$L_g(\psi) = g(y|\psi) = \frac{g(y,\theta|\psi)}{g(\theta|y,\psi)} = \frac{g(y|\theta,\psi)p(\theta|\psi)}{g(\theta|y,\psi)}, \tag{B.3}$$

so that

$$\frac{p(\theta|\psi)}{g(\theta|y,\psi)} = \frac{L_g(\psi)}{g(y|\theta,\psi)}.$$

Substitution into (B.2) gives

$$L(\psi) = L_g(\psi)\,\tilde{\mathrm{E}}\{\frac{p(y|\theta,\psi)}{g(y|\theta,\psi)}\}, \tag{B.4}$$

which is the convenient expression we will use in our calculations. The likelihood function of the approximating Gaussian model $L_g(\psi)$ can be calculated via the Kalman filter. The conditional density functions $p(y|\theta,\psi)$ and $g(y|\theta,\psi)$ can be easily computed given values for θ and ψ. It follows that the likelihood function of the SVX model is equivalent to the likelihood function of an approximating Gaussian model, multiplied by a correction term. This correction term only needs to be evaluated via simulation.

An obvious estimator for the likelihood of the SVX model is

$$\hat{L}(\psi) = L_g(\psi)\bar{w}, \qquad \bar{w} = \frac{1}{M}\sum_{i=1}^{M} w_i, \qquad w_i = \frac{p(y|\theta^i,\psi)}{g(y|\theta^i,\psi)}, \tag{B.5}$$

where θ^i denotes a draw from the importance density $g(\theta|y,\psi)$. In practice, we usually work with the log of the likelihood function to manage the magnitude of density values. The log transformation of $\hat{L}(\psi)$ introduces bias for which we can correct up to order $O(M^{-3/2})$; see Shephard and Pitt (1997) and Durbin and Koopman (1997). We obtain

$$\ln \hat{L}(\psi) = \ln L_g(\psi) + \ln \bar{w} + \frac{s_w^2}{2M\bar{w}^2}, \tag{B.6}$$

with $s_w^2 = (M-1)^{-1}\sum_{i=1}^{M}(w_i - \bar{w})^2$.

B.3 Computational Implementation

Given a particular vector for $\psi = (\phi, \sigma_\eta, \sigma_\varepsilon)'$, we evaluate the loglikelihood function (B.6) using importance sampling based on the Gaussian approximating model.

To obtain the maximum likelihood estimate of ψ, which we denote by $\hat{\psi}$, the loglikelihood is numerically maximised with respect to ψ. The repeated evaluation of the loglikelihood for different ψ's during the search for $\hat{\psi}$ will be based on the same set of random numbers used for simulation.

The loglikelihood functions of the SV class of models are estimated by Monte Carlo estimation. The SVX programs which are written in Ox 2.2 can be found on the Internet at `www.ssfpack.com/holbook`; also see Appendix C .

Appendix C
Data and Programs

This appendix provides an overview of the data series used in this book and lists the relevant software with which the various Stochastic Volatility models were estimated by the Monte Carlo estimation procedure. Both the data and the computer programs can be found on the Internet at `www.ssfpack.com/holbook`. The programs were developed together with Siem Jan Koopman. The required algorithms were implemented using the object-orientated Ox 2.2 programming language of Doornik (1998) and the state space library SsfPack 2.2 of Koopman, Shephard and Doornik (1999). The computer packages needed for running the programs can be downloaded from the Internet. Information about Ox can be found at `www.nuff.ox.ac.uk/users/doornik/` and further on-line documentation is provided at `www.ssfpack.com` and `www.feweb.vu.nl/koopman/sv`. The programs can be adjusted in order to use them in a more general context and for other Monte Carlo studies. In addition, they can be applied to other data-sets.

Chapter 2

Data	:	Financial Times All Share Index
Source	:	Datastream
Symbol	:	FTALLSH
Period	:	2 January 1975 to 30 September 2000
Frequency	:	daily, weekly and monthly
Program	:	SVmodel.ox

Chapter 3

Data	:	(i)	Financial Times All Share Index
		(ii)	Standard & Poor's 500 Composite Index
		(iii)	Tokyo Stock Exchange Index
		(iv)	UK Treasury Bill Discount 1 month - middle rate
		(v)	US 3 month Treasury Bill Rates - secondary market
		(vi)	Japan Bills 1 month - middle rate

Source	:	Datastream (i-iv, vi) Federal Reserve Statistical Release H.15 (v)
Symbol	:	FTALLSH, S&PCOMP, TOKYOSE, LDNTB1M, TB03, JAP1MBL
Period	:	2 January 1975 to 31 December 1998 (i-ii, iv-v) 4 January 1988 to 31 December 1998 (iii, vi)
Frequency	:	daily
Program	:	SVMmodel.ox

Chapter 4

Data	:	(i) Financial Times All Share Index (ii) Standard & Poor's 500 Composite Index (iii) Tokyo Stock Exchange Index (iv) DAX 30 Performance Stock Index (v) CAC 40 Stock Index (vi) Amsterdam AEX (EOE) Stock Index
Source	:	Datastream
Symbol	:	FTALLSH, S&PCOMP, TOKYOSE, DAXINDX, FRCAC40, AMSTEOE
Period	:	4 January 1988 to 31 December 1999
Frequency	:	daily
Program	:	SVmodel.ox

Chapter 6

Data	:	(i-a) Standard & Poor's 100 Stock Index - daily (i-b) Standard & Poor's 100 Stock Index - tick (ii) Chicago Board Options Exchange Market Volatility Index (VIX) - daily
Source	:	Datastream (i-a) Disktrading (i-b) Chicago Board Options Exchange (ii)
Symbol	:	S&P100I, $OEX, VIX(close)
Period	:	2 January 1986 to 29 June 2001 (i-a, ii) 6 January 1997 to 29 June 2001 (i-b)
Frequency	:	tick and daily
Program	:	SVmodel.ox, SVXmodel.ox, SVXplusmodel.ox, SIVmodel.ox

Chapter 7

Data	:	Standard & Poor's 100 Stock Index
Source	:	Disktrading
Symbol	:	$OEX
Period	:	6 January 1997 to 29 December 2000
Frequency	:	tick
Program	:	UCmodel.ox, SVmodel.ox, SVXmodel.ox

Bibliography

Akgiray, V. (1989), Conditional Heteroscedasticity in Time Series of Stock Returns: Evidence and Forecasts, *Journal of Business* **62**, 55–80.

Amin, K.I. and V.K. Ng (1997), Inferring Future Volatility from the Information in Implied Volatility in Eurodollar Options: A New Approach, *Review of Financial Studies* **10**, 333–367.

Andersen, T.G. and T. Bollerslev (1997), Intraday Periodicity and Volatility Persistence in Financial Markets, *Journal of Empirical Finance* **4**, 115–158.

Andersen, T.G. and T. Bollerslev (1998), Answering the Skeptics: Yes, Standard Volatility Models Do Provide Accurate Forecasts, *International Economic Review* **39**, 885–905.

Andersen, T.G., T. Bollerslev, F.X. Diebold, and H. Ebens (2001), The Distribution of Realized Stock Return Volatility, *Journal of Financial Economics* **61**, 43–76.

Andersen, T.G., T. Bollerslev, F.X. Diebold, and P. Labys (1999), (Understanding, Optimizing, Using and Forecasting) Realized Volatility and Correlation, Working Paper, Northwestern University, Duke University, New York University and University of Pennsylvania.

Andersen, T.G., T. Bollerslev, F.X. Diebold, and P. Labys (2001a), Applications and Case Studies - The Distribution of Exchange Rate Volatility, *Journal of the American Statistical Association* **96**, 42–55.

Andersen, T.G., T. Bollerslev, F.X. Diebold, and P. Labys (2001b), Modeling and Forecasting Realized Volatility, Working Paper, National Bureau of Economic Research, No. 8160.

Andersen, T.G., T. Bollerslev, and S. Lange (1999), Forecasting Financial Market Volatility: Sample Frequency vis-à-vis Forecast Horizon, *Journal of Empirical Finance* **6**, 457–477.

Areal, N.M.P.C. and S.J. Taylor (2002), The Realized Volatility of FTSE-100 Futures Prices, *Journal of Futures Markets* **22**, 627–648.

Bachelier, L. (1900), Theory of Speculation, In P. Cootner (Ed.), *The Randomness Character of Stock Market Prices*, Massachusetts Institute of Technology Press, Cambridge, MA, 1964.

Baillie, R.T. and T. Bollerslev (1989), The Message in Daily Exchange Rates: A Conditional-Variance Tale, *Journal of Business and Economic Statistics* **7**, 297–305.

Bakshi, G. and N. Kapadia (2001), Delta-Hedged Gains and the Negative Market Volatility Risk Premium, Working Paper, University of Maryland and University of Massachusetts-Amhert.

Barndorff-Nielsen, O.E. and N. Shephard (2001), Non-Gaussian OU Based Models and Some of Their Uses in Financial Economics (with discussion), *Journal of the Royal Statistical Society, Series B* **63**, 167–241.

Barndorff-Nielsen, O.E. and N. Shephard (2002), Econometric Analysis of Realised Volatility and Its Use in Estimating Stochastic Volatility Models, *Journal of the Royal Statistical Society, Series B* **64**, 253–280.

Beckers, S. (1981), Standard Deviations Implied in Options Prices as Predictors of Future Stock Price Variability, *Journal of Banking and Finance* **5**, 363–381.

Bekaert, G. and G. Wu (2000), Asymmetric Volatility and Risk in Equity Markets, *Review of Financial Studies* **13**, 1–42.

Bera, A.K. and M.L. Higgins (1993), ARCH Models: Properties, Estimation and Testing, *Journal of Economic Surveys* **7**, 305–366.

Black, F. (1976), Studies of Stock Price Volatility Changes, *Proceedings of the American Statistical Association, Business and Economic Statistics Section*, 177–181.

Black, F. and M. Scholes (1973), The Pricing of Options and Corporate Liabilities, *Journal of Political Economy* **81**, 637–659.

Blair, B.J., S. Poon, and S.J. Taylor (2001), Forecasting S&P 100 Volatility: The Incremental Information Content of Implied Volatilities and High Frequency Returns, *Journal of Econometrics* **105**, 5–26.

Bollerslev, T. (1986), Generalized Autoregressive Conditional Heteroskedasticity, *Journal of Econometrics* **31**, 307–327.

Bollerslev, T. (1987), A Conditionally Heteroskedastic Time Series Model for Speculative Prices and Rates of Return, *Review of Economics and Statistics* **69**, 542–547.

Bollerslev, T., R.Y. Chou, and K.F. Kroner (1992), ARCH Modeling in Finance: A Review of the Theory and Empirical Evidence, *Journal of Econometrics* **52**, 5–59.

Bollerslev, T., R.F. Engle, and D.B. Nelson (1994), ARCH Models, In R.F. Engle and D.L. McFadden (Eds.), *Handbook of Econometrics*, Volume 4, pp. 2959–3038, Elsevier Science, Amsterdam.

Bos, C.S., P.H. Franses, and M. Ooms (2002), Inflation, Forecast Intervals and Long Memory Regression Models, *International Journal of Forecasting* **18**, 243–264.

Brailsford, T.J. and R.W. Faff (1996), An Evaluation of Volatility Forecasting Techniques, *Journal of Banking and Finance* **20**, 419–438.

Brenner, M. and M.G. Subrahmanyam (1988), A Simple Formula to Compute the Implied Standard Deviation, *Financial Analysts Journal 44, September-October*(5), 80–83.

Brodsky, J. and C.M. Hurvich (1999), Multi-Step Forecasting for Long-Memory Processes, *Journal of Forecasting* **18**, 59–75.

Campbell, J.Y. and L. Hentschel (1992), No News is Good News, An Asymmetric Model of Changing Volatility in Stock Returns, *Journal of Financial Economics* **31**, 281–318.

Campbell, J.Y., A.W. Lo, and A.C. MacKinlay (1997), *The Econometrics of Financial Markets*, Princeton University Press, New Jersey.

Canina, L. and S. Figlewski (1993), The Informational Content of Implied Volatility, *Review of Financial Studies* **6**, 659–681.

Chesney, M. and L. Scott (1989), Pricing European Currency Options: A Comparison of the Modified Black-Scholes Model and a Random Variance Model, *Journal of Financial and Quantitative Analysis* **24**, 267–284.

Chib, S., F. Nardari, and N. Shephard (1998), Markov Chain Monte Carlo Methods for Generalized Stochastic Volatility Models, Discussion Paper, Nuffield College, Oxford, (forthcoming in: *Journal of Econometrics*).

Chiras, D.P. and S. Manaster (1978), The Information Content of Options Prices and a Test of Market Efficiency, *Journal of Financial Economics* **6**, 213–234.

Christensen, B.J. and N.R. Prabhala (1998), The Relation between Implied and Realized Volatility, *Journal of Financial Economics* **50**, 125–150.

Christie, A.A. (1982), The Stochastic Behavior of Common Stock Variances: Value, Leverage and Interest Rate Effects, *Journal of Financial Economics* **10**, 407–432.

Claessen, H. and S. Mittnik (1996), Forecasting Stock Market Volatility and the Informational Content of the DAX-Index Options Market, Working Paper, Christian Albrechts University, Kiel, (forthcoming in: *European Journal of Finance*).

Cox, J.C., S.A. Ross, and M. Rubinstein (1979), Option Pricing: A Simplified Approach, *Journal of Financial Economics* **7**, 229–263.

Danielsson, J. (1994), Stochastic Volatility in Asset Prices, Estimation with Simulated Maximum Likelihood, *Journal of Econometrics* **64**, 375–400.

Day, T.E. and C.M. Lewis (1988), The Behaviour of the Volatility Implicit in the Prices of Stock Index Options, *Journal of Financial Economics* **22**, 103–122.

Day, T.E. and C.M. Lewis (1992), Stock Market Volatility and the Information Content of Stock Index Options, *Journal of Econometrics* **52**, 267–287.

de Jong, P. and N. Shephard (1995), The Simulation Smoother for Time Series Models, *Biometrika* **82**, 339–350.

Diebold, F.X. and J.A. Lopez (1995), Modeling Volatility Dynamics, In K. Hoover (Ed.), *Macroeconometrics: Developments, Tensions and Prospects*, pp. 427–472, Kluwer Academic Publishers, Amsterdam.

Dimson, E. and P. Marsh (1990), Volatility Forecasting without Data-Snooping, *Journal of Banking and Finance* **14**, 399–421.

Donaldson, R.G. and M.J. Kamstra (2001), Volatility Forecasts, Trading Volume and the ARCH vs Option-Implied Tradeoff, Working Paper, University of British Columbia and Simon Fraser University.

Doornik, J.A. (1998), *Object-Oriented Matrix Programming using Ox 2.0*, Timberlake Consultants Ltd., London,
`http://www.nuff.ox.ac.uk/Users/Doornik/`.

Doornik, J.A. and M. Ooms (2001), A Package for Estimating, Forecasting and Simulating Arfima Models: Arfima package 1.01 for Ox, Working Paper, Nuffield College, Oxford.

Drost, F.C. and T.E. Nijman (1993), Temporal Aggregation of GARCH Processes, *Econometrica* **61**, 909–927.

Dunis, C.L. and J. Laws (2000), The Use of Market Data and Model Combination to Improve Forecast Accuracy, Working Paper, Liverpool Business School, (forthcoming in: *Applied Financial Economics*).

Durbin, J. and S.J. Koopman (1997), Monte Carlo Maximum Likelihood Estimation for Non-Gaussian State Space Models, *Biometrika* **84**, 669–684.

Durbin, J. and S.J. Koopman (2000), Time Series Analysis of Non-Gaussian Observations Based on State Space Models from both Classical and Bayesian Perspectives (with discussion), *Journal of the Royal Statistical Society, Series B* **62**, 3–56.

Durbin, J. and S.J. Koopman (2001), *Time Series Analysis by State Space Methods*, Oxford University Press, Oxford.

Durbin, J. and S.J. Koopman (2002), A Simple and Efficient Simulation Smoother for State Space Time Series Analysis, Discussion Paper, London School of Economics, Department of Statistics, (forthcoming in: *Biometrika*).

Ebens, H. (1999), Realized Stock Volatility, Working Paper, John Hopkins University.

Engle, R.F. (1982), Autoregressive Conditional Heteroscedasticity with Estimates of the Variance of United Kingdom Inflation, *Econometrica* **50**, 987–1008.

Engle, R.F. (Ed.) (1995), *ARCH: Selected Readings*, Advanced Texts in Econometrics, Oxford University Press, Oxford.

Engle, R.F., D.M. Lilien, and R.P. Robins (1987), Estimating Time-Varying Risk Premia in the Term Structure: The ARCH-M Model, *Econometrica* **55**, 391–407.

Engle, R.F. and V. Ng (1993), Measuring and Testing the Impact of News on Volatility, *Journal of Finance* **48**, 1749–1748.

Fama, E.F. (1965), The Behaviour of Stock Market Prices, *Journal of Business* **38**, 34–105.

Feinstein, S.P. (1989), Forecasting Stock-Market Volatility Using Options on Index Futures, *Economic Review (Federal Reserve Bank of Atlanta)* **May-June**, 12–30.

Feinstein, S.P. (1995), The Black-Scholes Formula is Nearly Linear in σ for At-the-Money Options; Therefore Implied Volatilities from At-the-Money Options are Virtually Unbiased, Working Paper, Boston University.

Figlewski, S. (1997), Forecasting Volatility, *Financial Markets, Institutions & Instruments* **6**, 1–88.

Fleming, J. (1998), The Quality of Market Volatility Forecast Implied by S&P 100 Index Option Prices, *Journal of Empirical Finance* **5**, 317–345.

Fleming, J., B. Ostdiek, and R.E. Whaley (1995), Predicting Stock Market Volatility: A New Measure, *Journal of Futures Markets* **15**, 265–302.

Fleming, J. and R.E. Whaley (1994), The Value of Wildcard Options, *Journal of Finance* **49**, 215–236.

Fletcher, R. (1987), *Practical Methods of Optimisation* (2 ed.)., John Wiley, New York.

Franses, P.H. and D. van Dijk (1996), Forecasting Stock Market Volatility Using (Non-Linear) GARCH Models, *Journal of Forecasting* **15**, 229–235.

French, K.R., G.W. Schwert, and R.F. Stambaugh (1987), Expected Stock Returns and Volatility, *Journal of Financial Economics* **19**, 3–29.

Fridman, M. and L. Harris (1998), A Maximum Likelihood Approach for Non-Gaussian Stochastic Volatility Models, *Journal of Business and Economic Statistics* **16**, 284–291.

Gallant, A.R., P.E. Rossi, and G. Tauchen (1992), Stock Prices and Volume, *Review of Financial Studies* **5**, 199–242.

Geske, R. (1979), A Note on an Analytical Valuation Formula for Unprotected American Call Options on Stocks with Known Dividends, *Journal of Financial Economics* **7**, 375–380.

Geweke, J. (1989), Bayesian Inference in Econometric Models using Monte Carlo Integration, *Econometrica* **57**, 1317–1339.

Ghysels, E., A.C. Harvey, and E. Renault (1996), Stochastic Volatility, In G.S. Maddala and C.R. Rao (Eds.), *Handbook of Statistics*, Volume 14, Statistical Methods in Finance, pp. 119–191, North-Holland, Amsterdam.

Glosten, L.R., R. Jagannathan, and D.E. Runkle (1993), On the Relation between the Expected Value and the Volatility of the Nominal Excess Return on Stocks, *Journal of Finance* **48**, 1779–1801.

Harvey, A.C. (1989), *Forecasting, Structural Time Series Models and the Kalman Filter*, Cambridge University Press, Cambridge.

Harvey, A.C. (1993), *Time Series Models* (2 ed.)., Harvester Wheatsheaf, Hemel Hempstead.

Harvey, A.C., E. Ruiz, and N. Shephard (1994), Multivariate Stochastic Variance Models, *Review of Economic Studies* **61**, 247–264.

Harvey, A.C. and N. Shephard (1996), Estimation of an Asymmetric Stochastic Volatility Model for Asset Returns, *Journal of Business and Economic Statistics* **14**, 429–434.

Harvey, C.R. and R.E. Whaley (1992), Dividends and S&P 100 Index Option Valuation, *Journal of Futures Markets 12*(2), 123–137.

He, C. and T. Teräsvirta (1999), Properties of Moments of a Family of GARCH Processes, *Journal of Econometrics* **92**, 173–192.

Hendry, D.F. and J.A. Doornik (1996), *Empirical Econometric Modelling Using Pc-Give for Windows*, Timberlake Consultants Ltd., London.

Heynen, R.C. (1995), *Essays on Derivatives Pricing Theory*, Ph. D. thesis, Erasmus Universiteit Rotterdam.

Hol, E. and S.J. Koopman (2000), Forecasting the Variability of Stock Index Returns with Stochastic Volatility Models and Implied Volatility, Discussion Paper, Tinbergen Institute, Amsterdam-Rotterdam, The Netherlands, `http://www.tinbergen.nl`.

Hull, J. and A. White (1987), The Pricing of Options on Assets with Stochastic Volatilities, *Journal of Finance* **42**, 281–300.

Jacquier, E., N.G. Polson, and P.E. Rossi (1994), Bayesian Analysis of Stochastic Volatility Models (with discussion), *Journal of Business and Economic Statistics* **12**, 371–417.

Jacquier, E., N.G. Polson, and P.E. Rossi (2001), An Extended Stochastic Volatility Model with Fait-Tails and Leverage Effect: Bayesian Analysis, Working Paper, Boston College, CIRANO and University of Chicago.

Jorion, P. (1995), Predicting Volatility in the Foreign Exchange Market, *Journal of Finance* **50**, 507–528.

Kim, S., N. Shephard, and S. Chib (1998), Stochastic Volatility: Likelihood Inference and Comparison with ARCH Models, *Review of Economic Studies* **65**, 361–393.

Koopman, S.J., N. Shephard, and J. Doornik (1999), Statistical Algorithms for Models in State Space using SsfPack 2.2, *Econometrics Journal* **2**, 113–166, `http://www.ssfpack.com`.

Koopman, S.J. and E. Hol Uspensky (2000), The Stochastic Volatility in Mean Model: Empirical Evidence from International Stock Markets, Discussion Paper, Tinbergen Institute, Amsterdam-Rotterdam, The Netherlands, (published in: *Journal of Applied Econometrics*, 2002), `http://www.tinbergen.nl`.

Koopman, S.J. and E. Hol Uspensky (2002), The Stochastic Volatility in Mean Model: Empirical Evidence from International Stock Markets, *Journal of Applied Econometrics* **17**, 667–689.

Kritzman, M. (1991), What Practitioners Need to Know about Estimating Volatility, Part 1, *Financial Analysts Journal 47, July-August*(4), 22–25.

Lamoureux, C.G. and W.D. Lastrapes (1993), Forecasting Stock-Return Variance: Toward an Understanding of Stochastic Implied Volatility, *Review of Financial Studies* **6**, 293–326.

Latané, H.A. and R.J. Rendleman (1976), Standard Deviations of Stock Price Ratios Implied in Option Prices, *Journal of Finance* **31**, 369–381.

Laurent, S. and J.P. Peters (2002), G@RCH 2.2: An Ox Package for Estimating and Forecasting Various ARCH Models, *Journal of Economic Surveys* **16**, 447–484.

Liesenfeld, R. and R.C. Jung (2000), Stochastic Volatility Models: Conditional Normality versus Heavy-Tailed Distributions, *Journal of Applied Econometrics* **15**, 137–160.

Lo, A.W. and A.C. MacKinlay (1999), *A Non-Random Walk Down Wall Street*, Princeton University Press, New Jersey.

Mandelbrot, B. (1963), The Variation of Certain Speculative Prices, *Journal of Business* **36**, 394–419.

Markowitz, H.M. (1952), Portfolio Selection, *Journal of Finance* **7**, 77–91.

Martens, M. (2001), Forecasting Daily Exchange Rate Volatility Using Intraday Returns, *Journal of International Money and Finance* **20**, 1–23.

Martens, M. (2002), Measuring and Forecasting S&P 500 Index-Futures Volatility Using High-Frequency Data, *Journal of Futures Markets* **22**, 497–518.

McMillan, D., A. Speight, and O. ap Gwilym (2000), Forecasting UK Stock Market Volatility, *Applied Financial Economics* **10**, 435–448.

Merton, R.C. (1973), Theory of Rational Option Pricing, *Bell Journal of Economics and Management Science* **4**, 141–183.

Nelson, D.B. (1988), *Time Series Behavior of Stock Market Volatility and Returns*, Ph. D. thesis, MIT.

Nelson, D.B. (1991), Conditional Heteroskedasticity in Asset Returns: A New Approach, *Econometrica* **59**, 347–370.

Nelson, D.B. and C.Q. Cao (1992), Inequality Constraints in the Univariate GARCH Model, *Journal of Business and Economic Statistics* **10**, 229–235.

Oomen, R.C.A. (2001), Using High Frequency Stock Market Index Data to Calculate, Model & Forecast Realized Return Variance, Working Paper, European University Institute.

Peters, J.P. (2001), Estimating and Forecasting Volatility of Stock Indices Using GARCH Models and (Skewed) Student-t Densities, Working Paper, Université de Liège.

Pitt, M.K. and N. Shephard (1999), Filtering via Simulation: Auxiliary Particle Filter, *Journal of the American Statistical Association* **94**, 590–599.

Poon, S. and C. Granger (2001a), Forecasting Financial Market Volatility, A Review, Working Paper, Strathclyde University and University of California.

Poon, S. and C. Granger (2001b), Practical Issues in Forecasting Financial Market Volatility, Working Paper, Strathclyde University and University of California.

Poon, S. and S.J. Taylor (1992), Stock Returns and Volatility: An Empirical Study of the UK Stock Market, *Journal of Banking and Finance* **16**, 37–59.

Quantitative Micro Software (1998), *EViews 3 User's Guide*, Quantitative Micro Software.

Ripley, B.D. (1987), *Stochastic Simulation*, Wiley, New York.

Roll, R. (1977), An Analytic Formula for Unprotected American Call Options on Stocks with Known Dividends, *Journal of Financial Economics* **5**, 251–258.

Rubinstein, M. (1985), Nonparametric Tests of Alternative Option Pricing Models Using All Reported Trades and Quotes on the 30 Most Active CBOE Option Classes from August 23, 1976 through August 31, 1978, *Journal of Finance* **40**, 455–480.

Sandmann, G. and S.J. Koopman (1998), Estimation of Stochastic Volatility Models via Monte Carlo Maximum Likelihood, *Journal of Econometrics* **87**, 271–301.

Schmalensee, R. and R. Trippi (1978), Common Stock Volatility Expectations Implied by Option Premia, *Journal of Finance* **33**, 129–147.

Schwert, G.W. (1989), Why Does Stock Market Volatility Change over Time?, *Journal of Finance* **44**, 1115–1153.

Scott, L.O. (1987), Option Pricing when the Variance Changes Randomly: Theory, Estimation and Application, *Journal of Financial and Quantitative Analysis* **22**, 419–438.

Shephard, N.G. (1996), Statistical Aspects of ARCH and Stochastic Volatility, In D.R. Cox, D.V. Hinkley, and O.E. Barndorff-Nielsen (Eds.), *Time Series Models in Econometrics, Finance and Other Fields*, Number 65 in Monographs on Statistics and Applied Probability, pp. 1–67, Chapman and Hall, London.

Shephard, N. and M.K. Pitt (1997), Likelihood Analysis of Non-Gaussian Measurement Time Series, *Biometrika* **84**, 653–667.

Sowell, F. (1992), Maximum Likelihood Estimation of Stationary Univariate Fractionally Integrated Time Series Models, *Journal of Econometrics* **53**, 165–188.

Stein, E.M. and J.C. Stein (1991), Stock Price Distributions with Stochastic Volatility: An Analytic Approach, *Review of Financial Studies* **4**, 727–752.

Stein, J. (1989), Overreactions in the Options Market, *Journal of Finance* **44**, 1011–1023.

Taylor, S.J. (1986), *Modelling Financial Time Series*, John Wiley and Sons, Chichester.

Taylor, S.J. (1994), Modeling Stochastic Volatility: A Review and Comparative Study, *Mathematical Finance* **4**, 183–204.

Taylor, S.J. and X. Xu (1997), The Incremental Volatility Information in One Million Foreign Exchange Quotations, *Journal of Empirical Finance* **4**, 317–340.

Tse, Y.K. and S.H. Tung (1992), Forecasting Volatility in the Singapore Stock Market, *Asian Pacific Journal of Management* **9**, 1–13.

Walsh, D.M. and G.Y. Tsou (1998), Forecasting Index Volatility: Sampling Interval and Non-Trading Effects, *Applied Financial Economics* **8**, 477–485.

Watanabe, T. (1999), A Non-Linear Filtering Approach to Stochastic Volatility Models with an Application to Daily Stock Returns, *Journal of Applied Econometrics* **14**, 101–121.

Whaley, R.E. (1981), On the Valuation of American Call Options on Stocks with Known Dividends, *Journal of Financial Economics* **9**, 207–211.

Whaley, R.E. (1982), Valuation of American Call Options on Dividend-Paying Stocks: Empirical Tests, *Journal of Financial Economics* **10**, 29–58.

Whaley, R.E. (1993), Derivatives on Market Volatility: Hedging Tools Long Overdue, *Journal of Derivatives* **1**, 71–84.

Wiggins, J.B. (1987), Option Values under Stochastic Volatility: Theory and Empirical Estimates, *Journal of Financial Economics* **19**, 351–382.

Xu, X. and S.J. Taylor (1994), The Term Structure of Volatility Implied by Foreign Exchange Options, *Journal of Financial and Quantitative Analysis* **29**, 57–74.

Xu, X. and S.J. Taylor (1995), Conditional Volatility and the Informational Efficiency of the PHLX Currency Options Markets, *Journal of Banking and Finance* **19**, 803–821.

Yu, J. (2000), Forecasting Volatility in the New Zealand Stock Market, Working Paper, University of Auckland, (forthcoming in: *Applied Financial Economics*).

Index

Printed in the United States
71160LV00003B/86